The Man Who Made Plants Write

Essays by
Jagadish Chandra Bose

Translated and with an Introduction by
Sumana Roy

YALE UNIVERSITY PRESS
New Haven & London

The essays in this book are from the collection *Abyakta* by
Jagadish Chandra Bose, originally published in Bengali in India in 1922.

Yale University Press books may be purchased in quantity for
educational, business, or promotional use. For information, please
email sales.press@yale.edu (U.S. office) or sales@yaleup.co.uk (U.K. office).

Set in Ivy Mode and Ivy Ora Type by Integrated Publishing Solutions.
Printed and bound by CPI Group (UK) Ltd, Croydon, CR0 4YY.

Library of Congress Control Number: 2025946503
ISBN 978-0-300-27840-8 (hardcover)

A catalogue record for this book is available from the British Library.

Authorized Representative in the EU: Easy Access System Europe, Mustamäe
tee 50, 10621 Tallinn, Estonia, gpsr.requests@easproject.com

10 9 8 7 6 5 4 3 2 1

Contents

Translator's Introduction

If a parent worries for their child, what do the childless worry about? This is not a question that has been put to me yet, but it is one that I have imagined. Every time you hear of a plane crash, you think of your own child. You are grateful that they were not on that plane. I read these words in *Reader's Digest* when I was in middle school, when I couldn't have known what it meant to be either—parent or child. A life with my parents, and later with friends and acquaintances, produced this shorthand: a parent was someone who worried for their children, the worries changing contours at different stages of life. But none of this might have come to me had I not met a similar strain of worry in the work of Jagadish Chandra Bose.

Jagadish Chandra Bose was a Bengali scientist who convinced a largely skeptical world that plants are living beings. I first encountered his work in an essay called "Ahoto Udbhid" ("The Injured Plant"), in which he records a plant's response to different kinds of stimuli. Passionate and defensive in his argument, Bose, groping for analogies that would convey the emotional and intellectual stature he discerned in plants, turns to schoolchildren:

> I hit the plant with a cane. The plant's growth reduced imme-
> diately. It took the plant more than half an hour to forget this
> injury. After that it began growing spontaneously as before. O
> cane-wielding schoolmaster, there is no doubt that some have
> become judges of the High Court because of being pulled by
> the ears by you. That boys would grow tall because of being
> struck by your cane is doubtful though. All kinds of injury
> stunt growth.

Who was this man who hit plants with a cane and imagined how it would feel to pull them by the ears?

Born to a Brahmo Samaj family in Munshiganj (what is now Bangladesh) in 1858, Jagadish Chandra studied in Calcutta's Hare School and St. Xavier's College, though he also spent time in Faridpur and Bardhaman, where his father, Bhagawan Chandra Bose, was deputy magistrate. Bhagawan insisted that his son study in a Bangla language school before he was forced to acquire the English language, and Jagadish Chandra would continue to feel an affection, attachment, and patriotism for his native tongue throughout his life. "The language that a man learns in his mother's arms is the language in which he expresses his happiness and sorrow," he writes in the preface to *Abyakta*, a collection of his essays published in 1922.

Jagadish Chandra went on to study the natural sciences at the University of Cambridge and University College London, following which he returned to teach at Presidency University in Calcutta. He would eventually establish the Bose Institute in

1917, where most of his research instruments are still preserved. But even as his international profile rose, his ambitions remained rooted in Bangla:

> About thirty years ago, I'd written a few of my scientific writings and other essays in English. I had started researching on electric waves and life, and that led to my involvement in several legal cases. The court for this is abroad, where arguments can only be made in the European languages.... Isn't this insulting to our national life, to the life of our *jaati*, our community? To redeem this, I have tried to establish a scientific court in this country. I might not live to see the fruit of this; the fate of scientific institutions is in the hands of god.

Folklore has it that Bose, having accidentally stepped on *Mimosa pudica* (*lajjabati* in Bangla, meaning "shy one" or "shameplant"), and, surprised by the folding of its leaves, decided to find out more about plant behavior, plants' responses and ways of communication. "Is there any possible relation between our own life and that of the plant world?" This question, which we find recurring throughout Jagadish Chandra's work, propelled much of his research. In a letter to his friend Rabindranath Tagore, after naming Western scientists who have refused to accept that plants are living beings, he writes:

> What I am doing is against accepted opinion. Just as cutting a tree from near its root leads to its fall, its rest on the earth, similarly with many old theories. Many things will need to be rewritten, and written afresh, and, for this, battles will have to be fought with old conventional thought. What I have

discovered gives me a lot of courage, but patience—patience patience. This virtue we lack.

Thirteen years later, two months before the start of World War I, Jagadish Bose gave a talk titled "Plant Autographs and Their Revelations" to the Royal Institution of Great Britain. The concept of autographs, scripts, and handwriting was of great significance to Bose's emotional and intellectual intuition about life, about what constitutes the living. For someone whose shorthand definition of life was the ability to respond, writing—script, the autograph—was a manifestation of that response. In script was an argument of life, the right to live. The talk begins with the science of handwriting:

> There are professors of sciences bordering on the mystical, who declare that they can discriminate the character and disposition of anyone simply by a careful observation of his handwriting. As to the authenticity of such claims, skepticism is permissible: but there is no doubt that one's handwriting may be modified profoundly by conditions physical and mental.

Bose recognizes the unreliability of such a science, but still adopts it to frame his discussion of plant behavior:

> Such, then, is the history that may be unfolded to the critical eye by the lines and curves of a human autograph. Under a placid exterior, there is also a hidden history in the life of the plant. Storm and sunshine, the warmth of summer and the

frost of winter, drought and rain, all these and many more
come and go about the plant. What coercion do they exercise
upon it? What subtle impress do they leave behind? Is it possible to make the plants write down their own autographs,
and thus reveal their hidden history?

What Bose seeks, therefore, is not "autograph" alone, but rather
the autobiography, the "hidden history," the "unsaid" (*abyakta*)
of the plants. Writing at the height of modernism, when autobiography was the entry point for those who had been kept out
of "literature" and "culture" by gatekeepers, this urge seems natural. The sciences can tell a story—of an animal, an object, an
element, a star, or, of course, a plant. Bose is signaling for a move
away from a literature *about* the other for a literature *by* the
other. He is challenging the Aristotelian hierarchy of "living beings" in which plants sink to the bottom. He is trying to parse a
language where none was said to exist. The vocabulary of this
challenge comes to him from a philosophy and everyday practice of enumerability. *Abyakta,* the title of his book, means the
unmanifest, and it is this that Bose wants to record through his
instruments. Here he relies on synecdoche, that a part will rep-
resent the whole of the plant's life history.

From time to time, people have reminded me that Jagadish
Bose and his wife, the feminist social activist Abala Bose, had no
children of their own. A colleague at a college where I once
taught, a professor of physics, was the first person to do so. "You
are like Jagadish Chandra Bose," he said to me, "a plant-mad parent who could not have children of his own." It was a lack, my
childlessness. Or not just a lack, it was a wrong.

I can't quite remember how the lack began, though I do remember the transition from "When will you have a child?" to "Why don't you have a child?" At some point later, it changed to "Because you don't have children . . . ," as in, because I didn't have children, I had become a "plant parent." I dislike that phrase, dislike its characterization, dislike how it alternates between consolation and moral superiority, how it turns plant life into an other, how parenting is turned into the likeness of a genre. But there it was: because I did not have children, I had started mothering plants. Had Jagadish Bose done the same?

He would go on to propose a *torulipi*, a plant script based on a record of a plant's responses to different kinds of stimuli, condensing a metaphor into a compound word. There was another such compound word that he was fond of using: *brikkhoshishu*, or plant-infant. In "Udbhider Jonmo O Mrityu" ("The Birth and Death of Plants") he writes, "Infants do not have teeth; they only drink milk. Plants too do not have teeth—that is why they can only partake of liquid and air." A few sentences later, he returns to the image: "Leaves have many tiny mouths. Seen closely with a microscope, all these mouths reveal tiny lips. When they no longer need food, the lips close." The metaphor occurs repeatedly throughout his work: "The seed hides under the earth for a long time. Months go by. Spring follows winter. Then the rains start—a day of rain. There is no need to hide anymore. It is as if someone is calling the child from outside, 'Don't sleep anymore, climb out now, you'll see the light of the sun.'" And: "Have you seen a seedling sprout out of the earth? It seems like a child is raising its tiny head to watch a new world with wonder." And: "Rejecting the darkness of the marginalized forests, the plant-child lifts its head."

As a metaphor, this would mean little had it not driven his curiosity to actually articulate a language of plants. We notice Bose defending his ambition to understand plant linguistics—if I may call it that—by referring to the plant-child equivalence:

Do plants say anything? Many will say, Now, what kind of a question is this? Have plants ever spoken? Can man express himself clearly? And what he cannot express, is it not language? We have a child—he cannot speak clearly; the few words that come out of him are so half-formed and even broken that it is impossible for anyone to understand their meaning. But we can understand everything that our child says. Not just that, though. There are many things that our little boy does not say aloud in words; his eyes, the movements of his face and hands, the shaking of his head—he speaks through these gestures; we understand that language as well, but others don't. One day a pigeon from a neighboring house came and sat on our house—it then began cooing and grunting at the top of its voice. Our little boy was thus introduced to the pigeon; he soon began imitating the pigeon's doob doob. How does the pigeon call? As soon as we asked him this, he would imitate the bird's call....

Returning home one day, I found the little boy with fever; a severe headache had made him lie limp on the bed. The naughty boy who prances around the house restlessly all day was now struggling to even open his eyes. I sat by his bed and ran my fingers through his hair. Recognizing me from my touch, he opened his eyes with a lot of effort and looked at me. He then made the pigeon sound. I heard many things in his pigeon call. I understood that the little boy was saying,

"You've come to see the little boy? The little boy loves you a lot." I understood several other things, things that I wouldn't be able to express in words. If you ask me how I could hear so many things in that pigeon call, there is only one answer: it is because I love the little boy. You have seen that by looking at her son's face a mother understands what he wants. Often there is no need for words. If one observes from love, many qualities are revealed; one is able to hear many things.

This is an uncommon scientific instinct and vocabulary: *love* ("if one observes from love, many qualities are revealed")—does one ever meet that word in scientific tracts? I've often paused at this passage, not just to remind myself of the behavior of the human mind before "critique" became the majoritarian position in the humanities, but from a desire to understand—or perhaps speculate on—what methodology such "love" and "observation" might point to, particularly when the research involved harm and cruelty to those whose behavior was being investigated. Bose employs his intuitive understanding of unity, of aesthetic, scientific, and mathematical truth, in his use of the word "love."

In his presidential address at the literary conference in Mymensingh in April 1911, which would later be published as "Literature and Science," Bose deliberates on the "caste system" in academia, the relationship between poetry and science, the "invisible" and the "unvoiced"—the first being light, the second being living forms whose language we do not understand. He then turns to communication between plants and humans, and how it is necessary for plants to write their own "script." "It is comparatively easy to make a rebellious child obey; to extort answers from plants is indeed a problem!" The words that follow are born of guilt, confessions of "various acts of cruelty" he has enacted in his research. "I have from time to time perpetrated violence on unoffending plants, in order to compel them to give me answers. For this purpose, I have devised various forms of torment—pinches, simple and revolving, pricks with needles, burns with acids. But let this pass. I now understand that replies so forced are unnatural, and of no value. Evidence so obtained is not to be trusted." This is the voice not of a scientist but of a parent—one who realizes, in retrospect, the guilt of having had their way.

I've often found myself distracted by the word *nursery*—a word for a space where both saplings and children are nurtured in their early life. It is hard to say, even when propelled by great investigative energy such as had possessed me, where Bose's emotional vocabulary, of plants as children of those without offspring, might have come from. The writer Dakshinaranjan Majumdar collected grandmothers' tales from villages and provinces in Bengal in the late nineteenth century and published them in 1907 with a preface by Rabindranath Tagore: "Sadly, this bag of mouth-watering tales is being supplanted by imports from England. The worst sufferers are the children. . . . Can a newborn baby who is fed barley water instead of nourishing mother's milk ever grow into a healthy child?" (It's the same anxiety about language we hear in Bose's thoughts about the use of Bangla in scientific argument; the child-mother metaphor is not an accident.) Through these Bengali tales of kings and queens, good and evil, and fantastic kingdoms runs a common motif—perhaps only visible to those who would have noticed it, for deprivation and accusations of lack make us notice things we might not have otherwise. The queen—the woman—must bear a child, and for this, different cures are invoked: worship, fasts, sacrifices, donations, rituals, austerity, piety, and, when everything else fails, a trust in the miraculous power of plants. A wandering yogi, passing through the kingdom of the childless king, would leave a root, a tuber, or some leaf for his many wives to grind into a paste and then share among themselves. Then the miracle would happen: the queens would all get pregnant. There are many variations on what followed after this: sometimes the women bear animals instead of humans, or half-humans, an early ancestor of a multispecies being; sometimes the newborn is kidnapped or smuggled out of the palace by jealous queens; and so on.

We find the same preoccupation in Tagore's own work. Bose, whom Tagore refers to as his "first friend," often visited the Tagore household. He taught Tagore's son, Rathindranath, to trace turtle footprints and search for their eggs. "He would make all of us dig pits in the sand and, with wet towels round our heads, lie down in them to sun-bathe," Rathindranath recalls. It was a friendship rich with the exchange of ideas: Tagore would read a new short story to Bose, and Bose would share the results from a new plant experiment. Rathindranath continues:

> Jagadish Chandra was at this time making experiments to compare the reactions on the Living and Non-living to different kinds of stimuli. He believed the results he had obtained with the help of the delicate instruments he had invented would revolutionize the current conceptions held by the scientists regarding the nature of life.... When Jagadish was satisfied that he had obtained sufficient convincing data to acquaint

the scientific world [with] his discoveries, he wanted to go to England to give actual demonstrations of his experiments to scientists in order to convince them of the truth of his deductions. Father approached the Maharaja of Tripura and was able to get from him sufficient money not only to enable Jagadish to go abroad but to fit up his laboratory with the equipment that he badly needed.

Every time you hear of a plane crash, you think of your own child. You are grateful that they were not on that plane. And every time a heat wave comes, I think of my plants, particularly the thin ones, the ones that look like children even after years in a pot, and I worry for them. Even if I had children, I think I would still worry about plants, those in my house, in my garden, and elsewhere. "Seeds are a plant's children," Bose writes. "All plants leave behind children before they die." I become a vulture, I orbit around this statement. Is it this that distinguishes plants from humans? It is possible that we are all parents, irrespective of our age. It is also possible that to care for others, even for those who do not look like us, cannot see us, or do not bleed when injured, is not necessarily the exclusive mark of a parent.

Bose writes about a plant-parent's sacrifice:

The plant nurtures its seeds with the juice in its body. It does not care for its own body anymore. It distributes everything of itself bit by bit to its children. The body which was alive and healthy even until a few days ago begins to wither. The wind would once shake its leaves with a *hoo hoo* sound. Now the drying plant cannot bear the pressure of the air. *Thhor thhor—* the plant trembles with just one slap of the wind. Branches break one by one. And one day the plant breaks, it falls to the ground.

This is how a plant sacrifices its life for its children and dies.

But what about Rathindranath and Bose and myself? Without children, is it possible for us to distribute everything of ourselves "bit by bit" to another? How are we to die? Should we be allowed to die at all, without this "sacrifice"? Bose and Rathindranath were allowed to die—but then, one investigated plant behavior, the other designed gardens. I, I'm still here.

And from here, I've watched how Jagadish Bose's intuition, which compelled him to design instruments and experiments that were once mocked by a colonial scientific establishment, is gradually being proved true a hundred years later. It was this desire, to see a thinker recognized for his revolutionary under-standing of life—life, and not plant life alone—that compelled me to translate this book.

This book, *Abyakta* ("The Unsaid" or "The Unmanifest"), was published in 1922, the same year as James Joyce's *Ulysses*, T. S. Eliot's *The Waste Land,* and books by Virginia Woolf and Kather-ine Mansfield. It was also the year Bertolt Brecht's first play was staged, not to mention when the world met Picasso's Cubism and Sinclair Lewis's Babbit. W. H. Auden, summarizing the age, said that the "climate" had changed. No list of books from 1922 includes Bose's book, of course. "Climate" is an interesting word to hear a hundred years later, when it has moved from being a noun to functioning as an adjective—from Auden's literary or intellectual "climate" to the doom of "climate change." Bose, however, didn't use the word. Its utilitarian meaning would have put him off, as would its summarizing instinct. His ambition was different: he would let the plants speak for themselves, he would be a facilitator, he would design instruments that would record their language. A hundred years later, we are returning to that moment. It's also pertinent to ask ourselves what might—and should—have happened had Bose's work found better circulation outside colonial Bengal, from where, in a laboratory deprived of support and infrastructure, he carried on with his experiments.

I would like to spend a little time with Bose's book so as to show similarities and divergences between his thoughts and those of his contemporaries. The year 1922 also saw the publication of Ryunosuke Akutagawa's story "In a Bamboo Grove," the main source of Akira Kurosawa's film *Rashomon;* in it was, besides other innovations, the emphatic philosophical truism that there are multiple points of view. Bose couldn't have read that story, but his philosophical and scientific investigations were driven by a similar belief in the multiplicity of perspectives to be found in the world, a rejection of the centrality of the human. What Akutagawa, and later Kurosawa, were seeking to show through narrative, the Indian scientist wanted to import to an understanding of plant life. That the language of plants was just different from ours, for Bose, simply reflected the nature of language of various life forms that seem incomprehensible to other species.

In *Abyakta* we encounter Bose's restlessness and passion, his affection for the neglected. In the world of science, with its performance of objectivity, in which emotions are suppressed and scientists expected to behave with the neutrality of the instruments they use, this passion is, of course, an exception. Reading the prehistory of emotions that attended his experiments, something not generally granted to us by scientists, we become aware of the scientist's nerves as he sets out to discover and prove the nervous system in plants. Such questions and anxieties become manifest in the first section of his essay "The Silent Life" (1922):

> The sapling, growing inch by inch, its growth invisible to the naked eye—how will I record its growth from moment to moment with a machine? Will an external injury affect the character and rate of its growth? Does giving it food—or depriving it of food—change it, and how long does it take for the change to begin? Will giving it medicine or poison cause it to change? Will it be possible for one poison to counteract the effect of another kind of poison in the plant? Can the amount of one poison undo the effect of another?
>
> If a tree responds to an external injury, how long will it take for this response to materialize? Does the duration of that feeling change in a different situation? Will it be possible to get the tree to write about this time? How does the injury reach the tree's inside? Does it have a nervous system? If it does, how is the speed of the nerve's excitement communicated? Do favorable circumstances cause an incremental change in its speed? Do adverse circumstances prevent that? Are there similarities between our nervous system and the plant's? Can the changes in speed be recorded by the plant itself? Do plants have muscles like there are in the human heart? In the end, when death comes to a tree, is it possible to record that moment of nirvana? And does the tree respond fiercely to that moment before falling into an everlasting sleep?
>
> Only a history of these different moments, captured by different instruments, will give us the true and uninterrupted history of plant life.

This is the language of affection, of imploration, and indeed, of activism. It was this desire to discover the plant's script that made Jagadish Bose design a variety of instruments, none of which seemed to satisfy him completely, for he was always trying to design another one, one more sophisticated and acknowledging of the invisible life of plants—the electro-optic analogue, the shielded lens antenna, resonant recorder, phytograph, plant phytograph, automatic phytograph, Bubler instrument, plant sphygmograph, and, of course, the best-known among them, the crescograph. The design of these instruments came from his living in both the physical and botanical sciences. His sole collaborator was a tinsmith by the name of Putiram Das. Convinced that to know about plant life we must go to the plants themselves, he designed his scientific instruments to be as "sensitive" as possible, a word appropriate as much for the efficiency of these instruments as it is for his empathy. What he wanted to record was response to various kinds of stimuli, of course, but if one notices carefully, this is actually a record of injury, of hurt, a First Information Report of the effect of burning or cold, electric voltage or darkness. He wanted plants to write their autobiography in their own script, "torulipi," the plant script. This must be one of the most non-anthropocentric practical philosophies imagined by a human.

Like most Bengalis I'd have forgotten about Bose after the General Knowledge–like manner in which we encountered the discoverers: Newton, Columbus, Einstein, those who had found some key to explaining what were, in exasperated shorthand, called the "mysteries of the universe." Bose wouldn't have figured in this list—but his discoveries and his deductions would have, moving from laboratory and print into almost folk-like status, so that a mix of pathetic fallacy and anthropomorphism would come to annotate the vocabulary of our daily consciousness—as when my mother would say, as if this was the most natural thought, that we shouldn't pluck a flower or fruit from a plant after sundown, for they were "sleeping." When I moved beyond the perimeter of my small town and its consciousness and met people who had not been raised in a similar culture—in Europe, for instance—a word could come at me, from empathy as much as from an awareness of foreignness: "animism"—that such thoughts as Bose's and my mother's were skin and muscle to an animistic culture. Like a fish who has no sense of being wet, we—and I include Bose in the collective pronoun—took such a conclusion to be natural, the only possible manner of thinking. That is why when I use the word "research" for Bose, I probably mean his intuition, for I think he replaced the experiment-observation-inference code with experiment-observation-intuition. His research—in metal fatigue, and in radio waves, his development of the Mercury Coherer Detector—like Newton's and Copernicus's, would change the way we saw and spoke about the world. He wanted to listen to them—plants and metals and waves—to record their response to the world, to life. And yet in other countries, particularly the United States, his work was rejected, and he became the subject of mocking cartoons, both in the US and in England and even in his own country, his Bengal.

A hundred years after the publication of this book, as scientists and humanists begin to acknowledge what they call "plant intelligence" and even "plant emotions," borrowing the nouns from a register of self-congratulatory humanism, we notice the lineage of Bose's unacknowledged work—the sensitivity of plants to sound and electric voltage, their ability to store and communicate information, that they, to use the words of the plant scientists in Tel Aviv University, "do not suffer in silence . . . when thirsty or stressed, they emit 'airborne sounds' in response to stress or when they are cut," that they "cry" or "can also be anesthetized and lose their responses to external stimuli . . . plants are known to produce endogenous anesthetic compounds to deal with stress." As you read this book, a collection of talks and essays delivered and written for children and adults, you will perhaps discover the revolutionary imagination that compelled the public scientist, a philosopher of life, to seek answers to the most fundamental question, one that I think possessed him in everything that he did, from his investigation of radio microwave optics to his work on the behavior of metals to the plant sciences: what is life, and is response the sign that marks the living?

Acknowledgments

I'm grateful to Jennifer Banks, who after reading two essays that I had shared with her, saw a book that she believed the English-language reader would—and should—want to read; to Agnibha Maity and Ecem Saricayir, who brought me some *aaram* during the editing process.

The Man Who
Made Plants Write

1

Folded Hands

The past constitutes the present; not knowing the history of the past will keep the present and the future unknown to us. To recover these old histories, a knowledge of how people spent their daily lives is necessary. From a folk tradition I'd heard that portrayals of life and living from one and a half thousand years ago can still be found in the cave temples of Ajanta. The journey has become much easier these days, but when I visited Ajanta many years ago, there were almost no proper paths, no roads. It was about a day's journey from the railway station. A bullock cart was our mode of transport. It was only after some difficulty that I reached Ajanta. Crossing the mountain stream that fell in the middle, I saw a series of caves carved out of the mountain; inside was the highest level of craftsmanship! On the walls and ceilings of these caves are paintings that haven't faded even after the passage of more than a thousand years. In the fresco of the royal court I saw that the ambassador of Persia had come to see the king. In another place was an intense martial image, with armed women soldiers, fighting. In another corner were two clouds, coming from opposite directions, attacking each other. Idols had erupted, swirling in vapor, fighting fiercely with one

another. Such conflict has been raging since the eve of civilization; it is ongoing and will continue into the future. Light and
darkness are rivals, as are knowledge and ignorance, dharma
and adharma, righteousness and sin. When Savita ascends in the
seven-horsed chariot from the ocean to the east, darkness is
vanquished, and it merges into the western sky.

There's one image where a prince is surveying the flow of
people from his palace. The grief of the diseased, the suffering
of the grieving, pierces his heart. How is this misery to be broken? Today he will leave his kingdom and wealth and set out in
quest of an answer.

Today is Makara Sankramana, the day of the Great Renun-
ciation.

Coming out of the cave temple, overcast in half-darkness, I
saw a sculpture of the Buddha carved into the mountainside,
peace spread on the face. His sadhana, the dedication of his pursuit, has opened a path of peace that is beyond *sukha-dukha,* happiness and sorrow.

As far as the eyes can see, there is no sign of the human. *Dhu
dhu*—it is like a desert. There is no bridge to cross the gap between past and present. What I had witnessed in the darkness
of the cave was like a kingdom of dreams. I returned home with
a restless heart.

A few years later, I was invited to an aristocratic public building. There were many paintings; looking at them absentmind-
edly, I was suddenly surprised by one of them. I had seen this
image before—the Buddha sculpture from the cave temple, with
peace on its face! There was something in the picture that I
hadn't noticed before. Below the statue is a child sleeping on
a stone; in front of the child its mother, begging for Buddha's

blessing for her son's well-being with folded hands. He who has taken on the burden of suffering of all living beings will rid the mother of her sorrow. I saw another scene in my mind's eye:

Gautama fasting under the Bodhi tree. Sujata's maternal heart swelling at this sight of poor Gautama. Gradually, little by little, a bridge of affection and compassion was built between the past and the present, and the gap between their times and ours bridged.

A foreigner next to me said—Look, the face of the god is cruel: on the one hand, the mother had so much interest and devotion, but the god's gaze was unmoving! How can you see compassion in the face of this innocent woman's stone statue?

Then I remembered the wise men who worshipped nature. Is the difference between these ignorant mothers and agnostic scientists really that vast?

Isn't nature cruel, too? How many can see compassion in its unchanging, iron-hard rules? When the Ananta Shakti Chakra is raging, who lifts the living beings that are crushed by it?

Not everyone can see maternal love in the heart of nature. What we see is only a reflection of our own mind. What the eyes graze on is only a pretext. Only in the calm and deep water can true reflection be seen. How can an immovable impression be reflected on the ever-agitated heart when it is like frightened water?

The sea assumes a peaceful form only at his command, at whose command the endless ocean is enraged by the wind. Who, then, will say that the heart of that innocent mother is not charmed by a peaceful touch?

What we cannot see, the meditative mother, overwhelmed by the birth of her son, can see. In front of her, the image of loving Jagat Janani, the goddess mother of the world, is revealed behind the stone statue. When looking at it, I felt like nectar was flowing from above, making the mother and child white and pure.

Every so often, an incompleteness comes and kidnaps from the world its beauty and vitality. The interplay of light and darkness, happiness and sorrow, are tumultuous because of incongruity, though a painting is incomplete without the combination of light and darkness. Images out of only light or darkness remain vague. Life is often lacking in beauty when compared to the scene I have described. A child or the raised folded hands of the woman, as in the painting, can change the whole scene. Light and shadow, happiness and indispensable sorrow, stand in their respective places. Then the rays from those two raised, joined—and folded—hands pierce the darkness and illuminate the whole scene.

1894

2

The Resonance and Conceivable World of the Sky

The visible world is composed of earth, water, fire, air, and sky. This can be taken as a metaphor. Behind the innumerable events in this world there are three reasons: the first is matter, the second energy, the third space or sky.

Matter exists in three forms: in the earthen form—the solid state; in fluid form—water; in gaseous form—as air. Inert matter is constantly vibrating with energy. This universe is oscillating in space. Its incredible energy revolves continuously in an eternal cycle. It is this force that allows the universe to travel in the boundless sky, rising and merging again and again.

Let us first see how energy is transferred from one place to the other.

Everyone has seen a signal pole in a railway station. When a rope tugs on a distant wooden block, the block waggles.

Energy can also be seen being transferred by other means. A ship sails over a river; the engine causes the water to break into waves that hit the riverbank repeatedly. The momentum of the waves carries the force of the machine far away.

The fingers of musicians that hit the strings cause similar vibrations. These vibrations create waves in the air. Sound is knowledge produced by the injury to airwaves.

Even without musical instruments, there can be melody. In the quivering wind-blown leaves, in falling droplets of water, in the waves on the beach there is *sur*, there is melody.

The shorter the strings of the sitar, the higher the pitch. When air vibrates 30,000 times per second, an extremely high pitch can be heard by the ear. If the string is further shortened, the sound will suddenly stop. The string will continue to vibrate, waves will be generated, but this high pitch will no longer produce *dhvani*—sound—in the ear.

Who would think that even though hundreds of sounds enter our ears, we can't hear them all? Endless music is sung everywhere, outside the house and inside it, but it's beyond our sense of hearing.

I have spoken about the vibration of inanimate matter and its resultant sound. Besides this, there are the waves that are constantly being generated in the sky. Just as a wave is first produced in a musical instrument by plucking it with the fingers, an electric wave is produced in the sky in a similar manner. We hear the waves in the air with our ears, we see the waves in the sky with our eyes.

We often cannot hear the waves in the air. We cannot always see the waves in the sky.

When two metallic spheres are connected to an electrical device, the spheres will become electrified repeatedly and, because of the electric force, waves will spread in all directions in the air. If the string is shortened, that is, if the spheres are made smaller, the pitch will rise.

Thousands of vibrations will occur every moment, and then from thousands to millions and from millions to many millions of vibrations will be generated.

Imagine a dark room where the air is being repeatedly struck by an invisible force. Nothing can be seen, only a deep sound penetrating the silence, piercing the ears. As the frequency of vibrations is increased, the pitch will rise to the higher seventh note. The earsplitting sound will cease eventually and turn into silence. After this, even if millions of waves hit the ear, we will not know anything of it.

Let electricity produce waves in the sky; millions of waves will rush around in all directions at every moment. We will remain unscathed even if we are immersed in this turbulent sea of waves. Let the pitch rise gradually. Billions and billions of waves will be generated every second; the dormant senses will suddenly awaken, and the body will feel the heat. As the pitch rises even higher, more and more and more waves are produced, streaks of light will invade through the darkness. A further increase in frequency will gradually fill the house with yellow, green, and blue light. If the pitch rises even higher after this, the eyes will become overwhelmed, and the rays of light will again become invisible. Even if the sky resonates with countless vibrations after that, we won't be able to perceive it in any sense.

Still we are entirely lost in this sea! We are deaf and blind! What can I see? What can I hear?

Nothing! We have set out on the ocean with a couple of broken compasses.

As I've said before, heat and light are only electrical vibrations of the sky. The vibration we feel through the skin is called heat; the vibration that excites the sense of sight is called light.

There are many vibrations in the sky that are utterly imperceptible to our senses.

Blind people have imagined different forms of the same animal by touching different parts of an elephant's body. We imagine energy similarly.

Not too long ago we used to think of magnetism, electricity, heat, and light as different energies. Now we understand these to be various forms of the same energy. Everyone is aware of the relationship between magnetism and electricity. That heat rays and light are caused by electrical vibrations in the sky has been proved recently.

These waves move through the sky at the same speed; they bounce back in the same form when they hit a metallic surface, and when they fall from the air onto another transparent substance, they twist in the same manner. The frequency of vibrations is the sole cause of the difference.

The sun is nine crore [93 million] miles away from this earth. The atmosphere extends up to 45 miles above us. After that there is *shunya,* space, nothing. For now, there is no connection between this earth and the distant sun.

But when the solar tidal wave rises in the fiery ocean, the earth becomes agitated by this solar upheaval—electrical currents begin to circulate around the entire planet.

So what we thought was separate is actually not separate. Billions of worlds scattered in the void are woven into the fabric of the sky. The vibration of one world is carried to another world.

The sun's rays falling on the earth take on various forms. Trees grow because of sunlight, flowers get their color. Vibrations in the sky, in the form of radiation, disturb the carbon molecules in the air to form the plant's body. Sun rays from many years ago are trapped in the tree trunk and buried in the earth. The released rays from coal illuminate roads and highways with gas and electric lights. This energy propels ships and steam engines.

Animals survive and grow by eating plants that have grown in sunlight. It can be observed that almost all motion of the earth's surface is caused by solar radiation. The vibrations in the sky are causing the earth to pulsate; the flow of life is in motion.

The veil was gradually lifted from our eyes. Now we can know that there are two reasons at the root of this multiform and diverse powerful world: the sky and its vibrations; and other inanimate objects.

Inanimate matter can be seen in various forms. Sometimes it appears solid like iron, sometimes liquid, sometimes gaseous, and sometimes in an even subtler form. Invisible vapor floating in the void and the rock-solid hard frost are the same substance, but how different they are in form and appearance!

The motionless air inside the house cannot be seen. Its presence cannot be immediately perceived in any sense. But when the invisible subtle air currents are set in motion, it acquires various properties. Everyone knows that a village can be destroyed in an instant by a harsh attack of the swirling invisible wind.

Inanimate matter constitutes the waves of the sky. Countless waves emerged from an unknown cosmic force at some point, leading to the creation of atoms. Their fusion brought the world and the universe together in the aggregation of numerous drops.

The world is floating in the sky-ocean, in its vortex.

The German poet Richter met archangels in the kingdom of dreams. The archangel said, "Human, you desire to behold the infinite creation of the Almighty—Come, you shall see the universe." Humanity, freed from earth's gravitational pull, embarked on a journey through the infinite sky with the angels. Piercing higher and higher, they began to move through the highest level in the sky. Before they knew it, they had left behind the seven planets and soon arrived in the solar region. The massive flame emerging from the fierce furnace of the sun did not burn them. They left the solar kingdom and arrived in a distant realm of stars. Counting grains of sand on a seashore may be possible for humans, but counting vast, scattered, and limitless worlds is beyond their imagination—to the south, to the left, to the front, to the rear, beyond the line of sight, the infinite numbers of countless worlds! Countless billions of suns circumambulate countless billions of planets; around them, countless billions of moons are orbiting—topless, bottomless, directionless, infinite! Later, they went beyond this cosmos to a more distant and unimaginable world. Suddenly their vision was obstructed by the unimaginable new universe covering all sides. Seeing the prime gathering of the unthinkable cosmos, humans were utterly exhausted and said, "Angel! Take my life away! May this body turn into unconscious dust. Let this body merge into unconscious dust. The unbearable weight of this infinity. Where is the end of this world?" Then the angel said, "There is no end before you. Are you tired of all this? Look back, and there is no beginning to this world either."

There is no end, no beginning.

The human mind cannot bear the weight of infinity. How can I conceive the idea of the boundless cosmos while being a mere speck of dust?

Through the microscope, one can see the vast universe in a tiny dot. Upon turning the perspective, the world transforms into a tiny particle, reversing the microscopic observation.

Leave the universe behind; focus your gaze on the smallest particle.

Inanimate matter is constantly assuming different forms in front of our eyes every moment. A great city disappears into the emptiness of flames. But that does not mean even a single particle is lost. A particle sometimes exists as soil, sometimes as plant, sometimes in a human body, and sometimes as invisible air again. No object is ever destroyed.

Energy is imperishable. A great energy surrounds the world; every particle is permeated by it. What I see at this moment I won't see exactly in the next. Just as a fast-flowing river continuously breaks and reshapes pebbles, this stream of mighty current is also constantly breaking and building the visible world. Since the beginning of creation, this stream has been flowing incessantly with an unstoppable momentum. It has no pause, no decrease, no increase. If there is an ebb in one part of the ocean, there's a flood in another part. Ebb and tidal flood are caused by the same reason. The quantity of water in the sea remains constant. As much as there is a decline in one place, there is growth in another place in the same amount. This ebb and flow—growth and decline—is moving around in all directions like waves.

Growth and depletion—the same pattern exists in the energy waves. Every object is constantly affected by these waves: the pebble is breaking and forming. Through this thrust of energy-projected waves, the world is kept alive.

Now let us leave the inanimate world behind to focus on animate life. Stirred by the touch of spring, waking the slumbering earth and filling the forests, plant-infants raise their heads from darkness. Lush landscapes blossom bit by bit in front of our eyes.

But where is the vitality, the life-enthusiasm of spring when autumn arrives? Flowers are nipped from the stalk, worn-out leaves lie fallen on the ground, and tree bodies are buried in the earth. Slumber comes after awakening.

Spring has returned; covered with dry petals and buried in seeds, the dormant plant-infants awake. The anticipation of death has made the tree store life-drops within its seeds. From those drops, the tree was reborn, it regained life.

It seems therefore that in every life there are two components. One is ageless, immortal; surrounding it is the mortal body. This bodily cover is left behind. The immortal life particle builds a new home with each rebirth. Part of that primitive life is passed down through generations to the present day! The vitality of millions of years ago still resides in the atoms of the flower bud that I am recklessly nipping from the stalk today.

Not only that. The inheritance of eternal life extends before every living being. The living being of the present time stands at the junction of eternity. Behind him is the history of the ages and an eternal future.

And humans? Before evolving into human form from the first speck of life, they have been through so many changes! Innumerable years, endless struggles, various strengths, and victory: life's highest achievement is the human!

The descendants of that microorganism, the weak creature, seek to forget their own imperfections and embrace boundless strength. They harness electricity from the sky for their own chariot. Even though he is ignorant and blind, the human is eager to recover the primitive history of the earth. He strives to look into the future by lifting the dense veil of darkness.

If ever a manifestation of divine power is possible in a living being, this is it.

What shall we call more astonishing? Which is more surprising—the infinitude of the universe or the attempt to grasp the infinite within this finite point?

I have said before that this world has neither a beginning nor an end. Now I see that there is neither small nor big in this world.

The pinnacle of life is the human being! This statement is not accurate for all time. The energy that has elevated the primordial speck of life to the human form, by whose exhilaration this multiform world and amazing life have arisen from a formless emptiness, is flowing evenly even today. Progress is in the upward direction of creation; and ahead lies an endless time and boundless advancement.

1894 (This essay was written for children)

3

The Plants' Story;
The Story of Plants

Do plants say anything? Many will say, Now, what kind of a question is this? Have plants ever spoken? Is it only humans from whose mouths words bloom? Can man express himself clearly? And what he cannot express, is it not language? We have a child—he cannot speak clearly; the few words that come out of him are so half-formed and broken that it is impossible for anyone to understand their meaning. But we can understand everything that our child says. Not just that, though. There are many things that our little boy does not say aloud in words; his eyes, the movements of his face and hands, the shaking of his head—he speaks through these gestures; we understand that language as well, but others don't. One day a pigeon from a neighboring house came and sat on our house—it then began cooing and grunting at the top of its voice. Our little boy was thus introduced to the pigeon; he soon began imitating the pigeon's doob doob. How does the pigeon call? As soon as we asked him this, he would imitate the bird's call. From then on, whether in happiness or sadness, always, coming or going, he makes the pigeon sounds. Learning the new art has brought him boundless joy.

Returning home one day, I found the little boy with fever; a severe headache had made him lie limp on the bed. The naughty boy who prances around the house restlessly all day was now struggling to even open his eyes. I sat by his bed and ran my fingers through his hair. Recognizing me from my touch, he opened his eyes with a lot of effort and looked at me. He then made the pigeon sound. I heard many things in his pigeon call. I understood that the little boy was saying, "You've come to see the little boy? The little boy loves you a lot." I understood several other things, things that I wouldn't be able to express in words. If you ask me how I could hear so many things in that pigeon call, there is only one answer: it is because I love the little boy. You have seen that by looking at her son's face a mother understands what he wants. Often there is no need for words. If one observes from love, many qualities are revealed, one is able to hear many things.

Earlier, when I would go to the fields or mountains by myself, I felt an emptiness. I then learned to love plants, trees, birds, worms, and insects. I am able to understand many of their words that I couldn't before. These trees that do not speak—that they have a life, that they eat like us, that they grow with time, I did not know any of this. I understand it now. I am able to see deprivation, sadness, and suffering in them now. To stay alive, they too have to be busy all the time. When fallen into bad times, some of them have to steal and rob as well. All the qualities that the human has, some of them are to be found in them as well. Plants help each other as well, they form friendships. Sacrifice, the human's greatest virtue—that, too, can be seen among plants. Plant life is only a shadow of human life. I will share this with you gradually.

All of you have seen dry branches of trees. Imagine that you are sitting under a tree. The tree is dense with green leaves, you are sitting in its shadow. A dry branch lies not far from the tree. Once upon a time, this branch had so many leaves, now they are all dry, termites are eating at its base. A few days more and there will be no trace of it. *Achchha,* tell me—what is the difference between this dry branch and the tree? The tree is growing, the dried branch is decaying; one has life, the other doesn't. What is living and alive will grow, gradually, continuously. The living has another trait—it is that they have speed; in other words, they move in all directions. One can't see a tree's movement—its speed—all of a sudden. Have you noticed how a creeper entwines itself around a tree? A living thing exhibits speed; a living thing keeps growing. Only in an egg can you not see signs of being alive. Life is asleep inside an egg. If it finds favorable heat, a baby bird is born. It is as if seeds are a tree's eggs; the tree's babies are sleeping inside its seeds. If the seed finds soil, heat, and water, an infant tree will be born.

There's a hard covering on the seed—that allows the baby tree to sleep deeply inside it. Seeds come in many forms and shapes—some are small, some big. It is not possible to guess the height of a tree from looking at its seed. A gigantic banyan tree is born from a very small seed. Who can imagine that such a large tree is hidden inside this tiny seed? You might have seen farmers scattering rice seeds in agricultural fields. But the plants and trees in the forests were not born from humans scattering seeds there. There are many ways in which seeds are dispersed. Birds eat fruits and carry their seeds to faraway lands. That is how trees come to exist on uninhabited islands. Besides this there is the air which helps seeds to travel to distant places. Many of you have seen a *shimool* tree. When the fruit of the *shimool* tree cracks open from the heat of the sun, its seeds begin to fly with the cotton inside it. In my childhood, we would run and chase these seeds; as soon as we reached out with our hands, the wind would push them up beyond our reach. This is how seeds are being dispersed day and night, through countries and continents.

Whether every seed will turn into a tree or not no one can say. The seed might fall on a hard rock—it won't be able to sprout into a sapling then. The seed needs heat, water, and soil to sprout. No matter where the seed might fall, the infant plant remains asleep inside it for many days. Until it gets an optimum place where it can grow, the hard cover of the seed will protect the infant plant from danger and difficulties.

Seeds ripen at different times. Mango and lichi seeds ripen in Baisakh (April–May); paddy and millets in Ashwin–Kartik (September–November). Imagine that the seed of a plant has ripened in September–October. There are storms toward the end of the month of Ashwin. Fallen leaves and branches lie scattered everywhere. That is how the seeds are dispersed as well. Who can tell where they are flown to by the strong winds? Imagine that a seed, after spending days and nights lying and being dragged on the earth, finds shelter under a brick or clod of earth. Where it was, where it has reached. It comes to be gradually covered with dust and soil. The seed is now out of the human's sight. It is true that it has gone out of our sight, but not out of God's range of vision. The earth picks her up into her lap like a mother. The infant plant, on being covered by earth, is protected from winter and storms. Safe, the plant infant lies asleep.

1894 (This essay was written for children)

4

The Birth and Death of Plants

The seeds hid under the soil. Month after month passed like this. After winter came spring. Then it rained for two days at the beginning of the monsoon. No need to hide anymore now. It was as if someone was calling the child from outside, "Do not sleep anymore; come up, you will see the light of the sun." The seed coat slowly fell off, and the sprout emerged from between two tender leaves. One part of the shoot went down and held on to the soil, the other broke through the soil. Have you seen seeds growing sprouts?

It seems as if the child is raising its little head to see a new country with wonder. The part that enters the soil when the plant sprouts is called the root. The other part that grows upward is called the stem. You will see these two parts, "root" and "trunk," in all trees. This is a thing of wonder; no matter how you place the tree, the root will go down and the trunk up.

There was a plant in a pot. To test it for an experiment, I hung the pot upside down for a few days. The head of the plant remained hanging down for a few days, and the roots went upward. After a couple of days, I saw that the plant, having realized it had been tricked, had grown alert. All its branches began bend-

ing to grow upward, and the root turned downward. Many of you would have spent time cutting radishes and making *shayata**
in the winter. You'd have noticed that at first the leaves and flowers are upside down. After a few days, one begins to see that the leaves and flowers have started growing upward.

The way we eat, the tree also eats. We have teeth so that we can eat hard things. Infants have no teeth; they can only drink milk. Plants also have no teeth, so they can only feed on liquid or absorb food from the air. Plants absorb sap from the soil through their roots. Sugar dissolves when water is added to it. Pouring water on the soil dissolves many things that reside inside the soil. Trees feed on those things. If the root of the plant is not watered, the plant's nutrition stops and the plant dies.

Even the tiniest of things can be observed with a microscope. Examining a root or the branch of a tree with this instrument shows that there are thousands of tubes inside a tree. Sap enters the plant's body from the soil through these tubes.

Besides this, the leaves of plants collect food from the air. Leaves have many tiny mouths. Seen closely with a microscope, all these mouths reveal tiny lips. When they no longer need food, the lips close. When we inhale and exhale, poisonous air is expelled by exhalation; it is called Angaraka Vayu, carbon dioxide. If it were to accumulate on Earth, all animals might die in a few days' time if they inhaled this poisonous air. Think about God's mercy, his karuna. What is poison for the animal, the tree, by feeding on it, cleans the air.

---

* The radish is cut in half and hung upside down on sticks. This is called *shayata*. When the plant has stored enough food, it turns from the bottom and grows upward, not toward the ground.

When the sun's rays fall on a tree, the leaves use the sun's energy to draw carbon out of the Angaraka Vayu. This carbon enters the body of the plant and helps it to grow. Plants need light; they cannot survive without it. The tree's primary endeavor lies in getting a little light. If you put the plant in a pot near the window, you will see that all its leaves turn to move toward the light. If you go to a forest, you'll see that the trees are trying to grow taller and faster so as to get the light first. If the climbers are in the shade, they will die from lack of light, so they cling to the tree and climb upward.

Now you can understand that light is the root of life. A plant grows by absorbing the sun into its body. The sun's rays are trapped in the body of the tree. The light and heat that comes out when wood is set on fire is the energy of the sun. Plants and their fruits, their seeds, are traps for catching light. Animals live by eating plants; the energy of the sun that is in the plants enters the animal's body in this manner. If we did not eat grain, we would not survive. If you think about it, we remain alive by eating light.

Some plants die after a year. All plants and trees are keen to have children before they die. Seeds are the offspring of plants. To protect the seeds, the plant prepares a small house with flower petals. How beautiful the trees look when they are covered in flowers. It seems that the tree is smiling. What else is as beautiful as flowers? Plants feed on food from the soil and on carbon from the air. How can such a beautiful flower be created from these little things? I have heard a story about a gem called *sparśamoni*, which turns iron into gold. I think a mother's love is that kind of gem. Love for the child blooms like a flower. It is as if a loving touch transforms the soil and the carbon into flowers.

How happy we feel when we see the flowers blooming on the trees! It seems like the tree is delighted as well! We invite ten people on a festive occasion, on a day of happiness. When the flower blooms, the tree also brings its friends. It is as if the tree invites with its call—"Where are my friends, come to my house today. If you forget the way and cannot recognize the house, I have made a sign of flowers of different colors. These colorful petals can be seen from far away."

Plants have everlasting friendships with bees and butterflies. They come in groups to see the flowers. Some insects cannot go out during the day from fear of birds. Birds eat them as soon as they see them. So they cannot go out unless it is night. To invite them, flowers release their fragrance in the evening.

Plants store nectar in their flowers. Bees and butterflies drink that nectar. Bees also help the plants. You might have seen the pollen in a flower. Bees carry pollen from one flower to another. Without pollen, seeds cannot ripen.

In this way the seeds ripen inside the flower. The plant nurtures its seeds with the juice in its body. It does not care for its own body anymore. It distributes everything of itself bit by bit to its children. The body which was alive and healthy even until a few days ago begins to wither. The wind would once shake its leaves with a *hoo hoo* sound. Now the drying plant cannot bear the pressure of the air. *Thhor thhor*—the plant trembles with just one slap of the wind. Branches break one by one. And one day the plant breaks, it falls to the ground.

This is how a plant sacrifices its life for its children and dies.

5

The Realization of Mantra

Many islands can be seen in the Pacific Ocean. These islands are made up of tiny coral skeletons. For thousands of years these invertebrates have been building these islands with their bodies.

All the impossible things that are being made possible by science today are also the fruit of the labor, even if small, of many people.

People were really helpless in the past. Intelligence, labor, and endurance have made man a king on earth today. What hardship and what labor man has put into achieving this progress and prosperity we can no longer remember. Who first taught us to light a fire, who first educated us about the use of metal, who invented the practice of writing—we know nothing about them. Whoever tried to introduce a new custom had to encounter obstacles at every step. Quite often they had to endure torture of various kinds as well. In spite of such hardship, they did not live to witness the success of their labor. In the present, it seems that the work has been in vain. But no effort is ever entirely futile. What appears to be extremely tiny today, excellent results emerge from it after a couple of days. Just as the coral island grows in size little by little, the kingdom of knowledge expands in the same manner. I will mention one or two incidents related to this.

A hundred years ago, Galvani, a professor in Italy, observed that touching a dead frog with an iron and copper wire made it move. He continued investigating this phenomenon for years. People used to mock him for wasting his time on such trivial matters, they began calling him the Frog Dancing Master. Friends would come and say, "The dead frog might move, but what good does it do?"

What is the benefit? That trivial incident inaugurated a line of new discoveries about the properties of electricity. How much of the history of our world has been changed by electricity in these hundred years. Streets are lit by electricity; cars are moving. In a moment news from one end of the world is reaching the other end. It is as if the whole world has come to a corner of our house—the distant is not far away anymore. Our voice wouldn't reach the other end of the house. Now, because of electricity, I have been speaking with a friend who is thousands of kilometers away. In fact, with the help of this energy, I can even see what is happening in far-off lands. Our voice and our vision will no longer obey any barrier.

Man has long dominated the earth and the sea, but he hasn't been able to conquer the sky. Airships can reach the sky, but balloons fail to move against the wind. Another disadvantage is that gas soon escapes from the balloon, so it cannot float anymore.

Since gas escapes through the silk covering, the balloon cannot remain in the sky for long. A German named Schwarz created an aluminum balloon for this reason. Aluminum is as light as paper, but gas cannot escape through it. No one would believe that a balloon could be made of metal. Schwarz exhausted his entire possessions on the experiment. After years of fruitless effort, the balloon was finally built. So that the balloon could go against the direction of the wind at will, he created a small engine. Just as a ship has a screw to propel it in water, a large screw was built so that the balloon could cut through the wind. But shortly after the balloon was created, Schwarz died suddenly. The thing for which he had risked his property and his life he could not test; his efforts were about to go in vain.

Schwarz's wife appealed to the German government to test the balloon. The German government was eager to use the aircraft in the war, but no one believed that Schwarz's balloon would ever take flight. The story of the widow's plight made the government depute a few superintendents from the war department to test the balloon. Many came to see the balloon on the designated day. The examiners found that the balloon was gigantic and, being made of metal, much heavier than a silk balloon. Then there were the engine and the many machines attached to the balloon to drive it. Could such a colossal thing ever fly in the sky? The examiners began discussing—such a strange machine will never be able to leave the ground. The man has died and his widow has come to show this with a lot of hope; a mere formality of examination will have to be conducted. There are several machines and gears attached to the balloon; cutting them to make the balloon slightly lighter might allow it to rise about four hands above the ground. Alas! He who created the balloon, his voice would no longer be heard on this earth. It took years to conceive all those machines that were removed as unnecessary. The machines that were removed as unnecessary had taken years to be created. They could have directed the balloon to move south, left, up, or down as desired.

There was one more hurdle after this. Who would steer the balloon in Schwarz's absence? Who else would understand the function of the machines? Whatever that be, an engineer from among the audience agreed to steer the machines as best as he could. From not too far away the widow counted every heartbeat of the machine. Would the balloon be able to leave the earth? The hopes of the deceased would be fulfilled this time or banished altogether. The machine was started; immediately the balloon left the earth and ascended into the sky at great speed. The wind was blowing at that time, but, ignoring its opposition, the balloon soared. Schwarz's efforts had succeeded at last. But the machines that people had considered unnecessary, their necessity was proven in almost no time. The balloon rose into the sky of course, but without those machines to control it, it fell to the ground, shattering into pieces. It was this adversity that led everyone to understand that the vision with which Schwarz had constructed the balloon might succeed someday. In ten years, it has indeed been proven. The airship that Zeppelin built became a formidable weapon in the war. This airship crossed the Atlantic Ocean with great ease after the war, and since then it has shortened the distance between Europe and America.

The airship needs to be filled with gas to become light, making it extremely large and expensive to build. Birds fly with such ease! Will humans ever be able to fly like birds? Large birds flap their wings a few times and, rising into the emptiness, spread their wings and move in circles in the sky. And, so moving, they disappear into the sky.

Have you never wished to fly like birds? In Germany, Lilienthal wondered why we cannot be tourists in the sky like birds. He then began experimenting, knowing that mastering this skill would take a long time. Just as a child learns to walk step by step with great effort, he too will have to learn to fly in the same way. If a child falls, they can try to get up again, but when one falls from the sky, there is no chance of rising again, death is inevitable. The knowledge of such risks did not prevent him from conducting his tests. Through his many experiments he prepared different kinds of wings, and, tying them to his arms, he jumped off mountains, and relying on the wings, he began descending. One day he felt that it was possible that the addition of more wings might make it easier to fly. It turned out that he was right. He conducted these experiments with great caution for thirty years. The greater part of his life had passed, so he was eager and enthusiastic about completing his work. The machine he had prepared now did not hold up as firmly as it had before. He tried to fly with this imperfect—this incomplete—machine. He was cutting through the air with quiet ease when, unfortunately, a sudden slap of wind broke an upper wing of the machine. He lost his life in this accident. But the new theories he discovered through those experiments now belong to the wealth of the world. His theories made the construction of flying machines possible later. In America, Professor Langley built a flying machine with wings; it had a very light engine. Many came to see it on the day it was tested. But the negligence of a mechanic caused a screw to come loose. The engine started and the machine took off, after which it began moving in circles in the sky. Then the loose screw came undone and the machine fell into the river. This failure left Langley with a broken heart, leading to his death.

The timid remain confined by the fear of trying and of death. The brave can transcend the fear of death. After Langley's death, his countryman Wilbur Wright started experimenting with flying machines. During one such flight, the machine stopped mid-air and Wright fell from the sky, breaking his leg. Undeterred, he started with his experiments all over again, and through his perseverance, man could become a sky-traveler and extend his empire into the blue sky.

6

The Invisible Light

The strings of a sitar vibrate when plucked, erupting into the resonance of *jhankar*. The string can be seen vibrating. That vibration generates invisible waves in the air, and through their impact, the melody is perceived by the auditory center. This is how, with the help of these three, news is sent and received between one place and another—the vibrating strings as the source of the sound; second, the transmitting air; and third, the cochlea, the auditory center as sound receiver.

The more the string of a sitar is shortened, the higher the pitch it reaches. When the air vibration reaches 30,000 times per second, an unbearable high-pitched sound can be heard. If the string is shortened further, the sound disappears. The string is still vibrating, but the auditory sense can no longer perceive that exceedingly high pitch. Just as there is an upper limit to our hearing range, there is also a lower limit. When a thick string or steel is struck, very slow vibrations can be seen, but no sound is heard. Vibrations between 16 and 30,000 per second are audible; our hearing is confined within this range, spanning about eleven sevenths, a gamut. So many sounds remain inaudible to us because of the limitations of our auditory sense.

Just as sound is produced by the vibration of air, light is produced by the vibration in the sky. Due to the limitations of our auditory sense, we can hear sounds within eleven sevenths. But the incompleteness of our visual senses is even greater; we are able to see only a seventh of the countless *sur* or frequencies in the sky. When the ether vibration reaches four hundred *lakh crore* [4 million million] times per second, the eye perceives it as red light; when the frequency of the vibration doubles, we see violet light. Yellow, green, and blue light—all fall within this one seventh. When the frequency of the vibration exceeds four hundred *lakh crore* times per second, the eye is defeated, and the scene fades into invisibility.

Whether visible or invisible, the vibration in the sky produces light. The question might arise: how can one detect this invisible ray, and what evidence do we have that this ray is indeed light? Let me describe an experiment on this subject. It was the German professor Hertz who first produced waves in the ether using electrical methods. His waves were so large that they did not travel in straight lines but instead curved. When a metal plate is placed in front of visible light rays, a shadow is cast behind it, but the large waves in the ether curve around the obstacle to reach behind it. This is similar to placing a stone in front of large waves of water. To prove that visible and invisible light share the same nature, the waves of invisible light need to be shortened. A machine I built produces waves in the ether that are only a sixth of an inch long. An electric wave is generated inside a small lantern. There's an open tube on one side through which the invisible light comes out. We cannot see this light; perhaps other creatures can. My experiments have shown that plants are stimulated by this light.

An artificial eye needs to be constructed to see invisible light. There is a retina made of nerves behind our eyes; when light falls on it, the flow of excitement travels through the nerves and stirs up a specific part of the brain. This stir we perceive as light. The construction of an artificial eye is quite similar. Two metal pieces are in contact with each other. When invisible light falls on the junction, a sudden molecular change occurs, resulting in the flow of an electric current that moves a magnet's needle. Just as a mute person signals by moving their hands, the artificial eye signals the perception of light by moving its needle when it detects invisible light.

The General Nature of Light

Let us now examine whether the nature of visible and invisible light is the same or different.

The nature of visible light is as follows:

1. It travels in a straight line.
2. When it falls on a metallic mirror, the light is reflected back. The reflection of the ray follows a specific rule.
3. The impact of light causes molecular changes. The natural properties of the illuminated material change. The image on a photographic plate undergoes chemical changes, and the picture becomes visible when the developer is poured on it.
4. Not all lights are of the same color; some light is red, some yellow, some green, and some blue. Different substances are transparent or opaque to different colors of light.
5. When light passes through air and falls into another transparent substance, it bends. This can be clearly seen when a ray or

beam of light is directed at a triangular prism. Light can be transmitted efficiently over a distance through a glass sphere.

6. Light waves usually lack any specific order; they are omnidirectional, meaning they can vibrate upward and downward or toward the left. By using a crystalline material, the vibrations of light rays can be chained, making the vibrations unidirectional instead of multidirectional. The special properties of unidirectional light I'll discuss later.

I will now describe the experiment that demonstrates that the nature of visible and invisible light is the same.

Firstly—the fact that invisible light travels in a straight path is proven by the fact that when the artificial eye is placed directly in line with the tube through which the electric wave exits the lantern, the needle moves. If the eye is placed to one side, no sign of excitation is observed.

Just as visible light reflects off a mirror, and this reflection follows specific rules, invisible light also reflects in the same manner and according to the same rules. Visible light causes molecular changes. I have been able to demonstrate through experiments that invisible light also causes molecular changes.

The Various Colors of Light

I have said above that visible light is of different colors; we can easily perceive the variety of colors through feeling, but there are many people who are unable to perceive any difference in colors. They are colorblind. I will discuss other ways to detect color differences later. Here it is important to say that the range of human vision is evolving. Our distant ancestors had a narrow sense of color, which has since expanded in at least one direction. It may expand in the other direction someday as well. When that happens, what is now left unseen will become visible.

In any case, I will describe a few strange experiments regarding the colors of invisible light. Window glass has no specific color; sunlight passes freely through it. So glass is transparent to visible light, as is water. But bricks are opaque, and tar is even more opaque. I've spoken about visible light. If the glass of the window is held in front of the invisible light, such light passes through it easily. However, when a glass of water is placed in front of it, the invisible light is entirely blocked. What a surprise that is! But there's something even more astonishing. Bricks, which we consider opaque, are transparent to invisible light. And tar? It is more transparent than window glass! I once read about a strange land where fish from ponds catch people by casting lines onto dry land.

In reality, though, it is not so. We have seen a similar astonishing phenomena with visible light, but we don't find them surprising because we are accustomed to them. On a piece of white paper in front of us here, two different light beams are falling, one red and one green. When a window glass is placed in the middle, both lights pass through freely. Now, if we place a red glass in the middle, the red light passes through easily, but the green light is blocked. If we place a green glass, the green light will pass through without obstruction, but the red light will be blocked. The reasons for this are: 1. Not all light is of the same color; 2. A material may be transparent to one type of light but opaque to another. Even without any knowledge of color, by observing that one light passes through a material while another doesn't, we can confidently say that the two lights are of different colors. Knowing that tar is opaque to visible light but transparent to invisible light proves that invisible light is of a different color. If our power of vision extended, we would be able to see new and unimaginable colors beyond the rainbow. But would even that satisfy our thirst for color?

Soil Sphere and Glass Sphere

I've said before that when light passes from one transparent substance to another, it gets refracted. This principle applies equally to visible and invisible light, as can be demonstrated using a triangular prism or a triangular brick. With the help of a glass sphere, visible light can be transmitted over long distances; invisible light can also be transmitted in a similar manner. An expensive glass sphere is unnecessary; a sphere can be constructed from bricks as well. I have successfully used the brick-built circular pillars in front of Presidency College to transmit invisible light over long distances. Diamonds have an extraordinary capacity to concentrate visible light. The greater a material's ability to concentrate light, the greater its ability to radiate it. This is the reason for the high value of diamond. It is astonishing that the capacity of Chinese porcelain to concentrate invisible light is many times greater than that of diamond. That is why if one day our vision extends beyond the limits of the red spectrum, diamonds will become insignificant, and the value of Chinese porcelain will increase immeasurably. When I first traveled to Britain, I was reluctant to touch Chinese porcelain, conditioned as I was to superstitions. When invited to their houses by well-reputed people, I saw Chinese plates and bowls displayed on the walls. What value did these hold to be treated with such care? I couldn't understand it initially, but now I realize that the British are merchants. If invisible light becomes visible, Chinese porcelain will become priceless. What is diamond compared to it, then? When that happens, fashionable ladies will proudly adorn themselves with garlands of cups and saucers instead of diamond necklaces, and they will look down upon unstylish women with disdain.

Omnidirectional and Unidirectional Light

The light from a lamp or the sun is omnidirectional; that is, the vibration is sometimes upward and sometimes downward and sometimes leftward. When light is passed through a tourmaline crystal from Sri Lanka, it becomes unidirectional. If two tourmaline crystals are held parallel to each other, light passes through both. But if one crystal is held at a right angle to the other, light cannot pass through both.

Invisible light can also be unidirectional in this way. To explain how this happens, one must recall the fable of the crane and the fox. The crane repeatedly invited the fox to drink a beverage. Beverages were stored in tall bottles. The crane, with its long beak, drank easily, but the fox could only lick the rim. The next day, the fox took its revenge. Drinks were served on a dish. The crane, despite tilting its beak, could not sip the beverage at all. Just as the difference between a long beak and a flat mouth can be understood through the bottle and the dish, the difference in unidirectional light can also be discerned—it is either vertical or horizontal.

Crane-Tortoise News

Imagine two groups of animals grazing in a field—the tall animal a crane and the flat animal a turtle. Omnidirectional invisible light is, similarly, caused by two types of vibrations. An easy way to distinguish between the two types of animals is to place a fence in front of them. When chased, the tall cranes will easily pass through the vertical railings, but the flat turtles will remain on the other side. If, after crossing the first barrier, a second fence is placed parallel to the first one, the cranes will still pass through it. However, if the second fence is placed so that the rails are horizontal, the cranes will be stopped. Similarly, when a single grid is placed in front of an invisible light, the light becomes unidirectional. If a second grid is placed parallel to the first, light will pass through it, making the second grid appear transparent to the light. However, if the second grid is rotated at a right angle, light will not pass through, making the grid seem opaque. So, if the light is unidirectional, certain objects will appear opaque when oriented in one direction but transparent when rotated by ninety degrees.

The pages of a book are arranged like a grid. During a lecture at the Royal Institution in Britain, there was a railway timetable on the table, a Bradshaw's guide, which contained the schedules, fares, and other information for ten thousand trains printed in tiny letters. It was so complex that it seemed impossible for anyone to extract necessary information from it. During my experiment, disregarding the darkness of the book, I showed that if you hold the book in a certain way, light cannot pass through it, but if you rotate the book by ninety degrees, it becomes completely transparent. The hall echoed with laughter when I demonstrated this experiment. I couldn't understand the reason for the laughter. Later, I understood. Lord Rayleigh came up and said that no one had ever seen light pass through a Bradshaw guide before, and that if I could teach people how to hold the book to see light through it, the world would remain eternally grateful to me. Some will be stunned, unable to gnash their teeth or roll their eyes after reading my scientific writing. In such a case, if you rotate the book by ninety degrees, all the information will become clear at once.

I was able to discover another way to make light unidirectional. Although disordered sky waves or vibrations enter a woman's hair, upon exiting, they become entirely organized. I collected various types of hair from the shops of hairdressers in Europe. The dense black hair of French women proved to be especially effective, the golden hair of German women much less so. When I demonstrated this experiment in Paris, the assembled French scholars were delighted by this new concept. Through this, their superiority over their rival nation has been proved; there was no doubt about it. Needless to say, I refrained from demonstrating this experiment in Berlin.

From the experiments described, it becomes evident that the nature of visible and invisible light is the same. It is due to the limitations of our vision that we perceive them as different.

Wireless News

Invisible light can easily penetrate through bricks and stones and buildings. So, with its help, news can be sent through the wireless. In 1895, I demonstrated various experiments related to this at the Calcutta Town Hall. The lieutenant governor of Bengal, Sir Alexander Mackenzie, was present. The electric waves

penetrated his large body and two closed rooms, creating various disturbances in the third room. It threw a metal sphere, fired
a pistol, and detonated a pile of gunpowder. In 1907, Marconi
received a patent for wireless transmission. His remarkable perseverance and contribution to the practical advancement of science have ushered in a new era. The distance between places on
earth has vanished completely. Only telegraph messages could
be sent to distant places before, but now wireless communication allows information to be transmitted everywhere.

Not only that. Human voices are now heard over great distances using wireless sky waves. Not everyone can hear these
voices; to listen, one's ear must be attuned to the frequencies of
the sky. Conversations continue day and night from one end
of the world to the other in this way. Listen carefully, opening
your ears. "Where is this message coming from?" The response
could be, "From under the sea, 300 feet deep. We have sunk
three warships with torpedoes and are waiting for two more."
What's that? The roar of millions of cannons can be heard as if
the earth itself is splitting apart from a volcanic eruption. Later
we learned that a great empire had been shattered and, from
tomorrow, the history of the world would be different. Amid
this dreadful noise one can hear many voices filled with deep
sorrow, so many pleas, many questions, and many different answers. In the midst of it all, someone was repeatedly calling out
a single name, like a bewildered person—"Where are you—where
are you?" No answer came—because that person was no longer
in this world.

The sound of the sky resonates far and wide in this manner.
Imagine an invisible finger striking various stops of an electric
organ. When it strikes the stop on the left, there is one pulse per

second. Immediately an electric wave races through the void. What a colossal wave, spanning thousands of miles! It effortlessly leaps over the Himalayas and circles the earth ten times a second. Now the invisible finger strikes the second stop. This time the sky pulses ten times per second. The pitch of the sky's sound will ascend higher and higher; the number of pulses will increase from one to ten, a hundred, thousand, million, and billion times. Immersed in the sea of the sky, we will be struck by countless waves, but no sense will be awakened. Let the sky's vibration rise higher; then, for a moment, heat will be felt. After that, the eyes will be stimulated, red, yellow, and other lights will be seen. This visible light is confined within the range of a *saptak,* a gamut. As the pitch rises further, the vision will be defeated again, sense and feeling will no longer awaken, there will be a moment of light, and then impenetrable darkness.

Then we are utterly lost in this infinity; how little can we see, after all? Insignificant indeed! We wander blindly within the infinite light and attempt to cross mountains with broken compasses. Oh, traveler on the infinite path, what is your possession?

There are no possessions but blind faith, the faith with which corals in the depths of the ocean are constructing great islands with their bodies of skeletons. The empire of knowledge grows bit by bit with such sacrifices of skeletons. It begins in darkness and it ends in darkness, only a few faint rays of light can be seen in between. Human perseverance will clear the thick fog and illuminate the entire universe one day.

7

Literature in Science

Various kinds of motion can be observed around the center in the physical world. The planets cannot escape the sun's gravitational pull; even an errant comet's trajectory eventually makes it run toward the sun.

Leaving the material world for the dynamic, the moving world, we notice how their movements seem quite irregular. Besides gravitation, there are countless other forces that constantly haunt them. They are constantly being injured, and their responses vary based on the nature and intensity of the impact, the hurt; they either laugh or cry in response. A soft touch and a gentle impact evoke physical excitement and the desire to get closer. But an increase in the intensity of impact causes them to change their impact accordingly. Where there was a gentle caress of the hand, there might be a sudden hit of a mace; where there was excitement and arousal, there might be terror and hesitation, a withdrawal. Repulsion instead of attraction, sorrow instead of joy, instead of laughter, tears.

The movement of loving beings is not solely determined by external impact. From inside various emotions appear and complicate external motion. How much of these internal emotions come from habit and how much from spontaneity? Can one determine the motion of a person driven by such diverse internal and external forces? But no one can resist the force of gravity. It is that invisible force that makes me return to my birthplace after so many years.

The attraction towards one's birthplace is natural because of one having come to be born here. But the logic behind my taking the position of the chair of this assembly today is not immediately evident. Questions may arise: is there place for a science worker in the field of literature? This literature conference has carried forth a deep and dense consciousness of the Bengali mind from one boundary of Bangladesh to another, awakening the striving for success everywhere. I can see clearly that there is no narrowness in the form and expression of the desire of Bengalis in this conference. Literature has not been confined within a tiny chamber or framework here; instead it seems that we have resolved to appreciate it more. Today literature is not just a beautiful ornament for us—today we are eager to see all the pursuits of our mind, all its sadhana, its aspiration, in literature.

In the *yajna* of this literature conference, I have also seen scientists. Among those honored with the position of the chair is our Desamanya Acharya Srijukta Prafulla Chandra, whom I consider a friend and colleague and, with pride and affection, refer to as a compatriot. By honoring him, the literature conference has not only worshipped virtue, it has also presented a liberal image of literature in our country.

There is an extreme reliance on specialization in the realm of knowledge in the Western world. Every branch of knowledge is organized to keep itself independent; as a result, the attempt to understand one's self has almost disappeared. In the initial stage of the pursuit of knowledge, this class-based segregation is helpful, facilitating, as it does, the collection and preparation of material. But if we only continue to follow this tradition until the very end, we fail to witness the entire essence of truth. Only the pursuit continues—we do not get any insight.

India, on the other hand, has always been careful to not lose sight of the one, the individual amid its multitude. As a result of this incessant pursuit, we are able to see it as one unit; it registers in our mind without any great resistance.

I can feel that this sense of unit has worked quite naturally regarding our literature conference. Right from the beginning of this conference, we have not confined the boundaries of literature to the narrow doors of authority. Instead, we have embarked on a path that has easily expanded its authority.

As a result, we are advancing toward a universal unity in the pursuit of knowledge without our awareness. We've simultaneously become eager to learn about our vast identity. What we desire, what we think, what we are examining—if we can find this in one place, we can see ourselves in true form. Those who have been singing, contemplating, and exploring all of them have, in our country, been invited to unite at this literary conference.

For this reason, even though I have spent most of my life in the pursuit of science, I do not feel any hesitation in accepting the invitation to this literary conference. For what joy could be more significant than embellishing what I have searched for, seen, and gained with the various treasures of our country? And if I have achieved the right to gather in a meeting with all the truth-seekers of our country today, what more joy could I possibly expect?

Poetry and Science

The poet, through his heart's gaze, perceives an incomparable entity in this universe, which he tries to express in form. Where the power of seeing ends for others, his vision of thought remains unhindered. Messages from that extraordinary country continue to emerge in the rhythm of his poetry in different addresses. A scientist's path might be independent and distinctive, but there is a similarity between the sadhana of the poet, the pursuit of the poetic, and his own path! He continues to follow the light even where the light of sight ends; even when the power of hearing reaches the limit of sound, he brings forth vibrating words. The scientist is trying to seek answers behind the mystery of illumination and carry it into a language comprehensible to the human.

The secrets of nature have countless homes, many mansions, infinite doorways. Natural scientists, chemists, and biologists have entered different houses through different gates; they thought of that particular quarter as their special place, and they cannot move to other realms. So, they have divided inert matter and plant consciousness into unbreachable partitions. But I do not accept that this division is a scientist's view. No matter how many walls are built for the sake of convenience of the quarters, each house has one overlord. All the sciences have journeyed through different paths with the same ultimate goal of discovering the truth. All paths have converged there, where the complete truth lies. Truth does not exist in fragments; it does not cause countless contradictions within. Therefore, we see biology, chemistry, and physics transcending their boundaries every day. The emotions of both the scientist and the poet are seeking an ineffable unity. The difference is that the poet does not think of a path, while the scientist cannot ignore a path. The poet must always be rapt; self-control is not for him. It is not for his poetry to provide evidence of his emotions; for this he has to turn to metaphor. To all his words "as if" needs to be added.

A scientist's path is quite a rough one; he must exercise self-control in this rigorous routine of observation and testing. He's always thinking: what if his heart betrays him? At every step he must align his inner thoughts with the external world. When the two sides do not converge, he cannot accept the perspective of one side alone.

Its reward is that although he cannot demand anything more than what he gets, he is assured of what he will get, and there is no weakening in the possibility of what he will be given in the future.

But such is the hard, uncertain journey the scientist has embarked on, all for its impenetrable mysteries. He has reached such a land of wonder where the barrier of tangible matter instantly dissolves in the face of invisible rays of light and where substance and energy merge into one. When the sight of an unimaginable kingdom suddenly bewilders the scientist and his blindfold is removed, he forgets his natural self-restraint and exclaims, "It seems as if it is, yet it is not."

Invisible Light

I'll invite you to enter an invisible world to witness an example of the close relationship between poetry and science. I shall say a word or two about whatever I have realized, part vaguely, part clearly, in a small corner of that boundless, mysterious realm. Even sighting the sea of light, with its crowd of colors, the poet's eyes remain unquenched. The seven colors cannot quench their thirst for sight. Is the boundary of this visible light crossed even while the boundless expanse of light continues to spread, then?

The German professor Hertz was the first person to show us this mystery. Numerous theories about the nature of this invisible light generated by electromagnetic waves have been discussed in the laboratories of the Presidency College. If I had the time, I could have demonstrated how this invisible light can detect the inner atomic composition of opaque matter. You'd then have noticed that there are many misconceptions about the transparency and opacity of matter. Light is freely passing through what we think is opaque. There are also strange objects which are transparent when viewed from one side and opaque when viewed from the other. You would have seen that just as visible light can be refracted far away by expensive glass lenses, similarly, invisible light can be refracted far away by earthen lenses. As a result, clay's ability to integrate invisible light is greater than the ability of a diamond to integrate visible light.

Only one of the seven notes of the countless gamut of the music of the sky excites our visual senses. That tiny perimeter is our visual realm. Amid boundless constellations we wander blindly. This incompleteness of humanity is unbearable. Yet the human mind never shatters completely; instead, with indomitable enthusiasm, in the raft of its incompleteness, it voyages through unknown seas in search of new land.

History of Plant Life

Just as there is an invisible light beyond the realm of visible light and, if we seek it out, our vision expands into the infinite, similarly when we grasp speechless pain beyond the realm of conscious states, our emotions can perceive the vastness of this area. Therefore, from the mysterious realm of Rudra's light, I invite you to the most resounding silence of the green plant kingdom.

Every day, as this vast world of plants unfolds before our eyes, is there any relationship between their lives and ours? Leading scholars in botany do not want to acknowledge any kinship with them. The famous Burdon-Sanderson says that, apart from a few, ordinary kinds of trees do not respond to external shocks, visibly or electrically. And even if sensitive plants respond electrically, their response is entirely different from that of an animal. Leading botanists in plant science, such as Pfeiffer, agree that plants are nerveless; plants have no such system as our nerve system that carries external messages.

From this it seems that different rules govern animal and plant life. The various problems in plant life are so complex that no fine-tuned machine has yet been discovered to understand it. This is primarily why people have relied on much-fabricated opinion instead of direct examination.

But if we want to know the actual theory, we have to leave the doctrine and try to get the direct result of the test. We must leave our imagination aside and question the tree itself, and rely only on the plant's handwritten description as evidence.

A Plant's Diurnal History

How will we know the internal changes in the tree? If the tree is excited due to some condition or depressed for some other reason, how can we understand these invisible changes from the outside? The only way is to capture and measure how the tree responds to all kinds of injuries from trauma.

When an organism is attacked by an external force, it reacts in various ways—if it has a voice, it cries out; if it is mute, it writhes its limbs. The reaction to external push or shaking is "response." We can measure the extent of life by correlating the measure of shaking with the measure of response. In an excited state, there is an intense response to minimal shaking. In a depressed state, there is a weak response to excessive shaking. And when death comes and overpowers living beings, all sorts of responses come to a sudden end.

The internal state of the plant might be inferred if I could record its responses on pen and paper by giving the tree some stimulus. If, somehow, I had succeeded in this seemingly impossible task, then we could learn this new script and language. Different countries have different languages; the letters used to write those languages are also different, and promoting a new script among them would be deplorable. The members of a script-society would feel offended, but there's no alternative. Fortunately, the writing of the plant is like Devanagari—utterly incomprehensible to the semiliterate or illiterate. Be that as it may, there are two obstacles to the fulfillment of purpose: the first of these is convincing the tree to testify about itself; second, recording that testimony with the help of both the tree and the machine. It is relatively easy to make a child obey commands, but obtaining answers from a plant is an extremely difficult proposition. At first, this attempt seems impossible. However, after years of closeness and forming a bond, I have understood their nature. Today, on this occasion, I confess to this kindhearted civilized society that I have behaved very cruelly toward innocent plants by forcing them to testify. For this, I have invented different types of pinches—straight or rotating. I have pierced them with needles and burned them with acid. I will not say more about that. But today, I know that the testimony that is obtained by this kind of coercion has no value; a fair judge can suspect this testimony as artificial.

If a plant could record its various responses with the help of its writing instrument, the actual history of the plant could be retrieved. But this is only a daydream. Such imagination merely causes slight excitement in the passive situations of our lives. The gratification of contemplation is easy, but, like opium, it gradually relaxes the gland.

When I get up from this state of dreaming and want to turn my imagination into action, I see an impenetrable wall before me. The temple of the nature goddess is made of iron. The demands and sobbing of a child do not reach past that door. But when a devotee with the power of accumulated intent over long periods breaks through the obstructed door, the nature goddess reveals herself to him.

Obstacles to Research in India

It is often heard that the lack of well-equipped laboratories has made research impossible in our country. Although this is largely true, it is not the complete truth. If it were true, new theories would be discovered daily from places where tens of millions of rupees have been spent building laboratories in other countries. But such news is never heard. We have many difficulties, we have many obstacles, this is true, but what is the use of envying the wealth of others? Get rid of your depression. Abandon weakness! Believe that whatever state we are in is our best state. India is our workplace; India is where we must fulfill our duties. He who has lost his manhood regrets in vain.

Apart from the lack of laboratories, there are other obstacles to conducting the tests. We often forget that the real laboratory is in our hearts. In that innermost country many tests are being conducted. Sadhana, this pursuit, is necessary to keep the insight bright. It fades very quickly. Where there is no impassive focus, external arrangements are of no use. Those whose minds rush outward, those who become greedy to get established in front of ten people instead of attaining the truth, do not get the vision of the truth. Those with no complete respect for truth cannot bear all sorrows with patience; their lust for quick fame makes them lose sight of the target. The path to success is not for those with such fickleness.

The lack of materials is not the primary deficiency for those who rightfully want the truth.

For the pure white lotus of Goddess Saraswati is not a golden lotus; it is a lotus of the heart.

A Plant's Writing

I was talking about the necessity of constructing fine instruments to record the various responses of trees. What was just a dream ten years ago has become a reality after years of labor. Now, there's no use saying how many attempts failed before there was success, and I will not exhaust your patience by describing the structure of these various machines. But it must be said that with the help of these different machines, the manifold responses of the tree will be recorded; the growth of the tree will be recorded from moment to moment. Its spontaneity will be recorded, and life and death lines will measure its existence. Suffice it to say that the amazing power of this machine is that calculation with it is so accurate that one thousandth of a second is easily determined. You will be pleased to hear one more thing. The construction of these machines has proved impossible in other more fortunate countries, but our artisans in this country have manufactured them. Its thinking and structure are entirely indigenous.

After many such experiments, I have been able to prove that the same rules govern tree life and human life.

In an article about twenty years ago I wrote, "Tree life is like the shadow of human life." I wrote it without knowing anything. I will confess that it was just youthful bravado and an excitement with words. Today, that lost memory has come back in words, and dreams and waking have merged together.

Conclusion

During this concluding session I shall tell you what I saw at the conference.

A long time ago, I visited a cave temple in the Deccan region. In the semidarkness of a cave, I saw Vishwakarma's idol standing. Various artisans were worshipping their tools at the feet of the deity.

As I watched this, I gradually understood that our arms are Vishwakarma's weapon. He uses this weapon to embellish the earth's lifeless clay in various ways. As a result of the advent of that great artist, our inert bodies have become conscious and creative. As a result of his advent, we, with the help of our minds and hands, have learned to establish the artist's various intentions in different forms: sometimes in art, sometimes in literature, sometimes in science.

The picture I saw in the cave temple is the same one I saw vividly here at the conference today. I have witnessed different tools of work within the Bengali mind that have been employed by our country's Vishwakarma. Somewhere, it is poetic imagination; somewhere, it is logic; and somewhere, it is the collection of information, of facts. We have placed all those ingredients, the materials, before him and come here to worship him.

The manifestation of this divine power in human power is the everlasting tradition of our country. It is divine power that causes creation and destruction in the world. Divine energy manifests in the human, for he can also create and destroy. The power to destroy our inertia, smallness, and failure is in us. The barriers of all these weaknesses are not an eternal truth.

The immortal are not born to remain small.

We also have the power to create. Our national greatness, which has almost disappeared, is still waiting for the creativity of our hearts. It is within our power to awaken the will and re-create it. The glory of our country, which once pierced the clouds, its speed, has not ended completely, it will touch the sky again.

It is an effort of our creativity that has taken a successful form in the Bengali Literature Committee today. We cannot just consider this committee a meeting place; its foundation is not laid on any exceptional roadside in Calcutta, and its building is not made of bricks. If we look inward, we will see that Sahitya Parishad exists as a temple before its devotees. Its foundation is established in the heart of Bangladesh, in Bengal; its building comprises layers—strata—of our lives. Before entering this temple, may we shun all the impure covering of our petty egoism and offer the purest flowers and fruits of our heart-garden as a gift of worship at the feet of the deity.

(President's address at the Mymensingh session of Bengali Sahitya Sammilani, a conference on Bengali literature, 1911)

8

The Speechless Life

As soon as I am out of the house, I see life—exploding. That life is completely soundless. Summer and winter, breeze and hurricane, rain and drought, light and darkness—they play with this silent life. The assembly of events, the many kinds of hurt, the variety of internal responses—inside this still-idol life an invisible sport is being played. How will I reveal this invisible to the world?

To know the real history of plants, we must go to the plants themselves. That history is mysterious and complicated. To retrieve that history, machines will have to record every moment of a plant's life, from birth to death. The script will have to be written by the plant itself, and it must have the plant's autograph. This is because men are often deceived by their motive.

The sapling, growing inch by inch, its growth invisible to the naked eye—how will I record its growth from moment to moment with a machine? Will an external injury affect the character and rate of its growth? Does giving it food—or depriving it of food—change it, and how long does it take for the change to begin? Will giving it medicine or poison cause it to change? Will it be possible for one poison to counteract the effect of another

kind of poison in the plant? Can any amount of one poison undo
the effect of another?

If a tree responds to an external injury, how long will it take
for this response to materialize?

Does the duration of that feeling change in a different situation? Will it be possible to get the tree to write about this time?
How does the injury reach the tree's inside? Does it have a nervous system? If it does, how is the speed of the nerve's excitement communicated? Do favorable circumstances cause an
incremental change in its speed? Do adverse circumstances prevent that? Are there similarities between our nervous system
and the plant's? Can the changes in speed be recorded by the
plant itself? Do plants have muscles like there are in the human
heart? In the end, when death comes to a tree, is it possible to record that moment of nirvana? And does the tree respond fiercely
to that moment before falling into an everlasting sleep?

Only a history of these different moments, captured by different instruments, will give us the true and uninterrupted history
of plant life.

Plant Script

Injury startles every living creature. That contraction is life's
response, the response of the living. When one is in the fullness
of life, the response is the highest in degree; in a moment of exhaustion, the response is feeble; there's none after death. There
is compression in a plant when it is hurt, but it is so little that we
are most often unable to see it. An instrument will amplify this
compression and record it as the plant's script. The only hurdle
for this is the extremely low intensity of the plant's response,

one that the instrument is often unable to script—its blade comes to a stop. To circumvent this, I was successful in inventing a flat-planed instrument. If two violins are tuned to exactly the same *sur,* if a string in one of them is stroked, the same string in the other will also create a *jhankar.* In the plant-script instrument, the iron wires inside and outside it are tuned to the same *sur.* Imagine this—that both the wires tremble one hundred times per second. If the external wire is played, the internal wire will move a hundred times, and the blade will draw a hundred dots. This is how the constant friction with the blade can be avoided. The transcript will record even the finest movement of the plant; this is because the difference between one dot and the next is only a second.

Plants—Are They Shy or Bold?

Before declaring the results of the experiment, it is important to dispel the superstition that surrounds plants—whether they are shy or bold, responsive or the opposite. That all plant life responds can be proven with the help of electrical energy. Why, then, is it that only *Mimosa pudica,* the touch-me-not, responds with its leaves when other plants do not? To understand this, please think of the muscles in your arm, by moving only some of which one can respond with one's hand. If the muscles on both sides of the arm were to move, then the hand would not be able to move. When a plant is hurt, all its muscles are compressed at the same time—that is why it does not move in any direction. But if a section of its muscles were to be anesthetized with chloroform, then it would be easy to prove the response and movement of plants.

The Latent Period

When a living being is hurt, it does not respond immediately. A frog takes a minimum of one hundredth of a second to respond if it is pinched on its legs. In English, this duration of time is called the "latent period."

This period increases or decreases depending on external circumstances. The latent period for a mild injury is somewhat long, but a severe injury is felt in no time at all. In winter, when the living being is slightly hesitant in moving, the latent period is long. When we are very exhausted, the period of response gets longer; sometimes, if the fatigue is acute, one might lose the power of response. The same is true of the latent period of the response of plants. The touch-me-not, when fresh, responds in one sixth of a second; the frog takes six times as long in its response. The strangest thing is that an obese tree responds to injury quite lazily. But the skinny rushes to the extremity of the seventh note immediately. Whether the same is analogically true of humans you might consider judging for yourself.

In winter, the latent period of plants grows twice as long. It takes up to fifteen minutes for the plant to return to its natural state. If it is hurt before that, the latent period multiplies one and a half times. If it is extremely tired, the sense of response goes missing temporarily, and the plant does not respond at all. What kind of state this might be you'll be able to see quite easily for yourself after my talk.

The Measurement of Response

The difference in time causes a difference in response. In the morning, the numbness of the night lingers, hence the plant's inertia. Repeated injuries cause this inertia to disappear, and its rate of response begins to increase; it is as if it were in a state of awakening. If the plant is given a hot water bath, the inertia will disappear quickly. In the afternoon, the opposite happens: tiredness causes its response to decrease. But if you allow the plant to rest, it gradually recovers from exhaustion. An increase in the impact of the injury will naturally tend to rev up the volume of the response as well, but this too is limited. Here, there is little to distinguish plant life from the human. Another surprising thing is that just as we take time to recover from our injuries in winter, so does the plant. What takes fifteen minutes to heal in summer takes more than half an hour in winter.

The Flow of Stimulus in Plants

The impact of injury in an animal's body is transferred through the nerves. The nervous system has a few characteristics: there is an increase in the speed of transmission at various times. Heat and cold affect its speed; electrical energy causes changes in the nervous system as well. As long as electrical energy flows through the nervous system, nothing much happens. But the transmission of this current just before it comes to a stop can cause excitement in one zone and exhaustion in another. The space through which the electrical current passes is suddenly excited. It does not allow for any "news"—or any new electrical impulse—to pass through the same space immediately. But once the electrical impulse is stopped, closed paths are reopened, and the nervous system becomes a news carrier once again.

With the help of instruments it is possible to prove and measure, in the finest detail, how nerves behave in a tree. The speed at which stimulus travels in a tree is much slower than how it travels in a frog; but it is faster than it is in lower animals. Heat causes the nervous system in a plant to double its response time. When electricity is administered to the tree, it causes excitement in one part of the tree and exhaustion in another. Electric current causes the push in its nervous system to stop. After having conducted all possible tests and experiments about the nervous system, I have been able to prove that there is no difference between the nervous system of plants and of other living creatures.

The Dance of *Desmodium gyrans*

It is possible to see the vibrations in the nervous system of the *Desmodium gyrans*. Its tiny leaves dance on their own. Some believe that the slightest clap can make them start dancing. Whether plants have any sense of music or not I cannot say, but there is certainly no relationship between the clapping of the human hand and the dance of the *Desmodium*. Having studied the script of a plant's vibrations, I can say with certainty that the nervous system in both plant and animal life works according to the same principle.

First, to ease the experiment, when the *Desmodium gyrans* is taken out of its pot, it stops vibrating. But if a tube is used to create pressure on the plant sap, it begins vibrating at an unregulated speed once again. Heat causes it to increase; cold reduces the rate of vibration. Ether stuns the plant's nervous system; a breeze helps it to come out of its unconscious state. The most interesting thing, however, is that just as poison causes the vibration of the heart to come to a stop, it behaves in exactly the same manner in a plant.

But I have learned to nullify the effect of one poison with another, even in a plant.

Now we need to find out the mystery of the vibration. My experiments tell me that an injury to a plant does not cause its muscles to respond immediately. But that does not mean that the power that enters the plant because of the impact of the injury is destroyed; the plant saves that power. This is how plants save the power that accrues from food, light, heat, and other sources of power. Only after the plant has hoarded as much as it needs does it let the leftover power flow out of it. That overflow we understand as the plant's vibration. What we think of as the plant's auto-vibration is actually an expression of its reserve of strength. When that reserve is denuded, the vibration stops. Pouring cold water on the *Desmodium* will stop its dance, its vibration. A little later, if it is able to find heat from its surroundings, the plant will start vibrating again.

Response to Death

In the end, a time comes in the life of the plant when a very strong blow or injury leads it to lose all capacity for response. That blow is the blow of death. But at that last moment the grace of the plant does not fade. Leaning toward falling or drying happens much later. When the harsh call of death comes, how does the plant respond? Just as a wave of regret passes through the human's body at the moment of death, we see a huge contraction of regret in the plant's body. At this time, an electric current passes swiftly through the body of the plant. The instrument recording the plant's script shows a sudden change of speed—the lines that were moving upward now run downward speedily, all before they stop. This response is the last response of the plant.

This is our voiceless friend, whose play of life continues near our doors. They have revealed their spirit and meaning in wordless letters in the plant script; their life's restlessness and their regrets about death have been published in front of our eyes. The difference that was created between the plant and animal world has now been dissolved. What was once news and beyond the imagination is now science—in clear scientific language we now see the unity in difference.

9

Thought and Action

Plant life is only a shadow, a reflection, of human life. I'd written an essay about this in a Bengali periodical. It didn't take more than an hour to write the essay. To establish a fraction of this has taken years, many years. To create the simplest substance and form requires great effort, perseverance, hard work, many many workshops and factories. But the mind is disobedient to boundaries. Thought can travel through heaven, earth, and hell in the blink of a moment. In an instant, concocted truths and falsehoods create new kingdoms, and if anyone obstructs the establishment of those kingdoms, then with the might of the mind and the help of the imaginary 218,700 warriors of the *akshauhini* armies of the Mahabharata and fiery arrows, the opposition is destroyed and a dictatorship is established.

But the customs of the workplace, of action, and the customs of the world of action are different—at every step there's an obstacle. I tried to govern my own life, but couldn't, with all my life's effort, and despite knowing this, the desire to assign others' duties while forgetting one's own never leaves. Giving up the more difficult path for what is easy and accessible, that which can be exhausted by speaking, desire naturally flows in that direction. Thought and action, what a difference there is between the two! The pace of action is even slower than the pace of a snail. The mind does not like to move along the difficult path of the kingdom of action. There is no lack of signposts, but where are the travelers on this path? In this bazaar of earthly existence, "everyone is a seller here, there are no buyers."

Everyone wants to see Baṅgajananī, the mother of Bengal, on a high throne, but needless to say, if we only persecute each other without hardship to ourself to find ways, nothing will bear fruit. For this it is essential for the children of Bengal to have a sense of self-respect; we forget this often. We only discuss others and the journeys they ought to make. Some have lamented that a few of Bengal's accomplished offspring have forsaken a noble path for trivial fame. Suffering from the same illusion, a few Bengali scientists could not resist the temptation of publishing their discoveries in foreign languages. If all these theories had been published only in the Bengali language, foreigners would have been forced to come to this country and learn Bangla from the attraction to invaluable truth; the West would have bowed its head to the East.

Speaking about the publication of scientific articles in English, it would suffice to say that whatever discoveries of mine have recently been established abroad were initially published in my mother tongue, and the experiments demonstrated before this country's general public. But such is my misfortune, I've not been able to gain recognition from the esteemed individuals of this country for a long time. When even the swadeshi universities of our country do not see any imprint of foreign influence, they become skeptical about the value of its truth. When these theories discovered in Bengal, written in Bangla, were rejected by Bengali scholars—nothing short of a tragedy—it is unreasonable to hope that foreigners would come to this country to search for precious gems amid garbage abandoned at the bottom of a river.

Only now have I come to understand that behind every barrier lies some intention. I have come to realize that the establishment of truth depends on adversity, for truth is weakened by the indulgence of favorable conditions. If scientific truth cannot be triumphantly brought through all the enemy territories like a sacrificial horse in an Aśvamedha Yajna, then the *yajna* ritual will not be considered complete. That is why I do not consider it my duty to take pride in pursuing truth in life; my goal is to help it conquer. A vast battlefield of scientific truth is spreading in the Western world today. Just as Bengalis had a notorious reputation on the battlefield once, Indians were often criticized in the field of science. Despite my repeated efforts to combat this, I've been defeated. I'd thought that my sadhana, my pursuit, would only result in failure in this life. But even amid deep despair, I've not accepted the disgrace of defeat. I crossed the Mediterranean for the third time and have, by the grace of the Almighty, achieved significant success. If I have gained victory laurels in this long journey, then I dedicate them to the feet of the goddess of this *desh*, this country, Desalakshmi. It is true that a few foreigners have tried to understand the immortal theories of our ancestors with great difficulty. However, I do not see any signs that the West has bowed its head to the East in the present age. Greek and European civilizations owe much to ancient Egyptian civilization, yet the descendants of Egyptian peasants are despised today. O descendants of the composers of the Vedas and the Upanishads, O farmers of India, where is your place today?

O, Alnaskara, will your daydreams never end? Your goods are only gilt and glass. You thought you would sell them as gold and diamonds, and you have trampled the goddess of fortune by thinking you are rich with your illusory wealth! Have you already forgotten the ridicule by spectators within a few days? What are you saying? Your ancestors were wealthy; they used to frolic in flying chariots! Fool! Then how did you lose that wealth? Look around. The white mountains you see in the distance are made of human skeletons. These are the skeletons of those whom you consider barbarians. Look, those who climb the stairs made of bones are rising to the summit of the mountain and, with a leap into the void, are extending their dominion into the azure sky. Who you considered to be the flock of eagles disappeared behind the clouds. Astonished, you gaze upward. Suddenly, a thunderbolt descends from the kingdom of clouds and shatters the world around you. Where will you seek refuge? There is no salvation, no escape, even if you enter the cave. You have to get out of that place because of the poisonous fumes.

In this entanglement of words we have spread a net, and you are trapped in it. You'll have to cut that net and come out. Do not bring your mother to the assembly without good reason. Her true place is in the temple of the heart; the sacrifice of life is necessary in order to worship her.

10

Offering

Twenty-two years ago I experienced the presence of a divine blessing in my life. I am offering what I desired that day to the divine feet after this long period of time. What I have established today is a temple, a *vidya mandir*, a temple of learning, not just a laboratory. The truth perceived by the senses is determined by examination, but beyond the senses there lie one or two greater truths; to attain these, one must rest only on faith.

Scientific truth is revealed through examination. That requires great practice. What was in the realm of imagination has to be made perceptible to the senses. The light that was invisible to the eye must be made visible. When the body-bound senses are transcended, one takes refuge in the metallic, in the elements that exist beyond the senses. The world that moments ago was so silent and dark now suddenly appears roaring and unbearably illuminated, and overwhelms us completely.

Although not all of these can be immediately perceived by the senses, they can be apprehended through artificially created human-made senses; but there are many more occurrences that are imperceptible to those senses, and that can only be realized through the power of belief. There is also an examination of the truth of belief, which does not happen through a single event but requires lifelong practice. It is for the establishment of that truth that temples are built.

What is the ultimate truth for which this temple was established? It is this: when a person dedicates their life and worship to a purpose, that purpose never fails; even the impossible becomes possible. Listening to the words of the wise is not my goal today, but my words are particularly for those who have leapt into the ocean of action and risen as lifeless corpses amid adverse currents, accepting defeat as fate.

Experiment

It took two lifetimes to complete the experiment I will talk about. As the real truth about all plant life is discovered through the experiment on a little plant, so is the truth of the realm of faith established by a belief in human life. Please overlook personal matters for the two words I will say about the truth tested in my own life. The experimentation started with my father, the late Bhagawan Chandra Bose, half a century ago. It is from him that I received my education and training. He taught me that governing one's own life multiplies benefits over seeking dominance over others, and dedicated his own life to the welfare of others. He directed all his efforts and resources toward the advancement of education, art, and commerce. All his efforts, however, ended in failure. From the soft bed of comfort and prosperity, he had to experience the harsh reality of poverty. Everyone used to say that he had failed in his life. It was from this event that I learned how success could be trivial and failure so immensely significant. The first chapter of the experimentation was written at this time.

Thirty-two years have passed since then; I have been engaged in teaching. While explaining the history of science, I had to remind students of the names of intellectuals from different countries. But where does India stand among them? In the realm of education, I had to teach everything that others had said. Indians are sentimental and dreamy; research work is not for them—this is one phrase I've heard forever. There are no laboratories in this country comparable to those in the West, and the construction of delicate and efficient instruments might never happen here. This, too, I have heard countless times. Then it occurred to me that the person who loses manliness is the only one who regrets it, and for no good reason. Despondency has to be dispelled and weakness renounced. India is our *karm-abhoomi,* our place of work, of action; an easy path is not for us. Thirty-two years ago on this day, keeping this in mind, a man dedicated his entire mind, soul, and sadhana to the future. He had no wealth, no guide. For many years he struggled alone, every day, against adverse circumstances. After all these years, his offering seems meaningful.

Victory and Defeat

Twenty-three years ago, on a dark day, the work I had started with hope of divine compassion bore fruit within three months. The extensive expansion and result of the difficult work on electromagnetic waves initiated by Professor Hertz in Germany was made possible right here. But when I shared the news of my discovery at a prestigious conference in this country, not a single dignitary expressed any opinion about my work! I realized that they are doubtful of an Indian's scientific achievements. I subsequently submitted my second discovery to the foremost physicists of the present time. Today, after twenty-two years, I have received their response. From this, I became aware that my discovery would be published by the Royal Society, and since this information will assist in future scientific advancements, a scholarship was granted to me by Parliament for my research work. On that day, the closed doors that were standing before India suddenly opened. No one can block those open doors anymore. The fire that was ignited on that day will never be extinguished.

With this hope I advanced in the field, year after year, with an unwavering mind and body. Human trials are not completed in a day; one undergoes repeated tests amid hope and despair throughout one's life. When my scientific aspirations were gaining an unexpected position, almost all aspects of life seemed to be on the verge of failure.

I was testing a machine that I was constructing to catch wireless news. The machine suddenly stopped responding for reasons unknown. I observed patterns in the machine's response script—it was similar to human scribble, and seemed to reflect a human's physical fatigue and weakness. What was also astonishing was that after a rest, the fatigue of the machine disappeared and it began responding again. The energy to respond increased with the application of a stimulating medicine, and poison brought its response to an instant halt. The power to respond, considered a significant characteristic of life, was observed in the behavior of inanimate matter as well. I was able to demonstrate this remarkable incident to the Royal Society through experiments. A few leading biologists were extremely displeased with it since it opposed prevailing opinion. I'm a physicist; I had attempted to venture unauthorizedly into the realm of biology by abandoning boundaries. This was against convention. One or two more unfortunate incidents followed. Someone who was against me claimed my discovery as his own. It is unnecessary to dwell further on this matter. Consequently, for many years, all my efforts went almost in vain. All these years I have not seen the face of light piercing the veil of clouds even for a day. These memories are extremely painful. The only reason for mentioning this is to underscore that if someone embarks on a great endeavour in life, they should remain neutral to the outcome. If boundless patience exists, then one day, with unwavering faith, you will see that he who, despite repeated failures, does not turn away ultimately wins.

World Tour

Fate and the cycle of action are spinning with rules—rise, fall, and rise again. The dense dark days of the past twelve years, which made me feel dejected but could not completely destroy me, that disaster also passed unexpectedly. To promote my new discovery in the scientific community, the Indian government sent me on a world tour in 1914. I presented my research in places like London, Oxford, Cambridge, Paris, Vienna, Harvard, New York, Washington, Philadelphia, Chicago, California, Tokyo, and others. In all these places, no one waited to receive me with laurels; instead, my staunch opponents were united in pointing out my flaws. I was completely alone then; the goddess of the fate of India alone was my invisible support. In this incomparable struggle, it was India's victory, and those who were once my adversaries later became my closest friends.

Code of Valor

The extraordinary accomplishments in the field of current plant science result from half a century of exceptional achievement by the German professor Pfeiffer of Leipzig. Some of my discoveries contradict Pfeiffer's opinions. Thinking I had incurred his displeasure, I did not go to Leipzig but accepted an invitation to Vienna University. There, Pfeiffer sent his colleague to invite me. He sent me a message that my newly established theories had reached him in the evening of his life; it was his regret that he could not see the fruition of these truths in his lifetime. The one who I feared would be hostile welcomed me as a friend. This is the eternal code of heroes where one is filled with joy seeing the victory of truth even amid one's own defeat. Three thousand years ago, this heroic duty was preached on the battlefield of Kurukshetra. When the arrow struck Bhishma's vital spot and he fell, he exclaimed in joy, "My teaching has borne fruit! This arrow does not belong to Shikhandi; it is my favorite disciple Arjuna's!"

Traveling the world and the tests of my own life have made me understand that discovering and pursuing new truths are essential. The dissemination of these in the world is even more difficult. This strengthens my old resolve. May the place that India has been able to occupy after a long struggle in the field of science be permanent. I hope that those who follow my work will never be deterred from their path.

India's Contribution to the Promotion of Science

Science is, of course, universal, but is there any place in the vast field of science that would be incomplete without Indian practitioners? Surely there must be. In our time the expansion of science has become widespread, and for the convenience of work in the West, it has become extensively divided, and unbreachable boundaries have emerged among various branches. The visual world is extremely diverse and has many faces. It is incomprehensible that there should be any commonality in such diversity. There is no resemblance between ever-fickle animals and eternally silent, impassive plants. That is, responses in various forms can be observed among these plants. However, amid such diversity, the system of Indian thinking has found a bridge of unity linking inanimate matter, plants, and animals. Indian practitioners sometimes freely send their thoughts to the realm of the unfettered imagination and immediately bring them under governance. They have infused new life into almost inanimate fingers by ordering it and have created an artificial transcendental sense where human senses fail. With this, and with boundless patience, they have dared to establish stability in the research methods of limitless mysteries of the unspoken world. What was once invisible to the eyes is now perceptible.

Examining artificial eyes, they have discovered a new mystery, imperceptible to human sight, where two eyes do not remain awake simultaneously; one sleeps while the other stays awake. The invisible imprint of hidden memories has been revealed in metal sheets. With the help of invisible light, they have revealed the inner construction of black stone. Molecular artifacts are demonstrated by rotary electrophoresis, through swirling currents of electricity.

By showing the portrait of human life in plant life, they have included the excitement of silent life to human feelings. They have measured the invisible intelligence of trees and monitored the changes in that intelligence in various stages of nourishment and usage. They have demonstrated that trees shrink upon human touch. The stimulant that excites man, the drug that depresses him, and the poison that destroys his life have been proven to have the same effect on plants. They revive plants that have withered due to poison through the application of a different drug. They recorded the pulsation of the plant muscle and showed that it reflected the heartbeat. They discovered the flow of nerves in the bodies of trees and determined their speed. They have proved that plant nerves are excited or subdued by the exact causes that increase or depress excitability in human nerves. All these words are not constructs of the human imagination.

This is a very brief and incomplete history of all the research that has been examined and proven over the past twenty-three years in my laboratory. Coming via different paths, various disciplines—physics, botany, zoology, even psychology—converge into one center in my research. If God has prescribed any special pilgrimage of science for the Indian practitioner, it is this *chaturveni sangam,* the confluence of two or more rivers, that is the great pilgrimage site.

Hope and Faith

All this research, this investigation, concerns various branches of science. Some believe that through their development, diverse practical advancements and the welfare of the world will be achieved. Will this hope and faith with which I established the temple cease to exist with the end of someone's life? Unlimited resources are necessary to construct an observatory for a single subject, and the spread of such broad and multifaceted knowledge is impossible for our country; any wise person will say that. But I've been walking a lifelong journey based solely on faith in this improbable subject. I have never turned away, saying, "This is impossible." I will not do that now. I will employ what I thought was my own in this work. I came with empty hands, and I shall leave with empty hands; if anything has been accomplished, I shall consider it the grace of God. I will employ my everything in this work whose company has remained steady amid my sorrows and defeats for many days. I have never been completely deprived of God's compassion. Even when many doubted my scientific prowess, the belief of a few surrounded me. Today, they are on the other side of death.

I feared that the stability of Vigyan Mandir, this temple of knowledge, would solely depend on the uncertain provisions of the future. A few days ago, I came to realize that the invitation of the hope with which I began the work has resonated even in the distant corners of India. Seeing all this, I feel that the outcome of the grand will I had conceived is not entirely impossible. Perhaps, during my lifetime, I may witness that the empty courtyard of this temple will be filled by travelers from near and far.

Discovery and Promotion

The practice of science has two sides. First is the discovery of new theories, which is the primary purpose of this temple. Then there's the promotion of these new theories to the world. It is for this reason that this large auditorium has been built. A lecture hall of this kind, dedicated to scientific lectures and their demonstration, has not been built anywhere else. It can accommodate one thousand five hundred listeners. There will be no repetition of extensively discussed theories here. All the new truths about science discovered in this temple will be disseminated with a demonstration here first. The doors of this temple will always remain open for men and women of every nation. Through the journal published by the temple, newly discovered scientific theories will be advertised to scholars worldwide, making the advancement of applied science possible.

My intention is that even those living abroad should not be deprived of the education provided by this temple. Centuries ago, knowledge was spread universally in India. In Nalanda and Taxila, students from abroad were welcomed in this country. Whenever we have had the power to give, we have given generously; we've never been satisfied with little. Our lives are enriched through contact with all living beings. What is true and what is beautiful are our deities. Artists have adorned this temple with their craftsmanship, painters have expressed the unspoken aspirations of our hearts on their canvases.

The plant life I have spoken of is an echo of our own lives. This life gets wounded and faints and wakes up again from a momentary stupor. There are two sides to this impact; we exist at the intersection of these two. On one side lies the path of life; on the other, the path of death. The vibrations of life are reactions to impacts from which we can rise again. Every moment we are wounded and brought to the brink of death, only to be revived once more. It is through these impacts that the strength of life is increased. Because we are dying bit by bit, we are alive, living.

A day will come when the impact will be overwhelming; what falls then will not rise again, and no one will be able to lift it.

Then the tears of loved ones will be in vain and the lifelong vows and worship of a devoted wife will be futile. But which realm does that death, the touch of which calms all anxiety and agitation, encompass? Who will unveil its mystery? We are utterly shrouded in the darkness of ignorance. Once the veil is lifted from our eyes, we become overwhelmed by the infinite expanse of an unimaginable new world beyond this small one.

Who would have thought that in this silent plant world, in this silent, infinite biome, a sense of emotions is developing? And from this nervous excitement, how did its shadowy, incorporeal tenderness and affection emerge? Which among these is ageless, and which is immortal? When the play of these animated puppets comes to an end, and their remains disappear into the five primary elements, will those incorporeal shadows fade into the sky or blossom in a greater form?

Over which realms, then, does death hold dominion? If death is the ultimate end of human beings, what will it do with the earth, so abundant in wealth and prosperity? But death is not all-conquering; its dominion is limited only to the physical realm, an aggregation of inert matter. The celestial fire born from human thought is not extinguished by the blow of death. The seed of immortality lies in thought, not in wealth. Great empires have never been established through conquest, but through the dissemination of thought and divine knowledge. Twenty-two centuries ago, the great empire that Ashoka founded in this very land of India was not established solely through physical strength and worldly wealth. What was accumulated in that great empire was meant only for distribution, for the alleviation of suffering and for the welfare of all beings. By distributing everything for the liberation of the world, a time came when the lord of sea-girt earth, Ashoka, was left with just half an amalaka fruit. Holding it in his hand, he said: Now this is my all; let this be accepted as my ultimate gift.

Offering

This symbol of the amalaka fruit is carved on the temple walls. At the very top, like a flag, the symbol of the *vajra* is established, an indestructible weapon made from the bones of the virtuous sage Dadhichi. The *vajra* is created from the bones of those who sacrifice their lives for others, and by its burning radiance demonic forces are destroyed, and divinity in the world is established. Today, our offering is only this half-amalaka fruit, but the glory of the past will undoubtedly be reborn in an even greater form. With this hope, we stand here for a brief moment; from tomorrow onward, we will once again set our life's boat afloat in the stream of action. Today, we have come here with offerings for the revered goddess. Her true abode is not outside but in the temple of the heart. Her true offering is found in the strength of the devotee's arms, the power within, and the devotion of the heart. What blessings will the devotee seek after that? When the devotee's pursuit does not end even after offering a burning life, an illuminated life, when he lies defeated and dying, waiting for death, then the revered goddess will take him into her embrace. Through such a defeat, he will receive his reward.

(On the occasion of the establishment of
Vigyan Mandir, or the Bose Institute, 1917)

11

Initiation

We are all learners, we're always learning at the workplace, we are progressing and growing with each day.

One of the great truths about life is that from the day our desire to grow is suspended, the shadow of death begins to fall on life. The same is true of national life. From the day the desire to grow ceased, our downfall began. We have to live, we have to save for our future, we have to grow. One needs to have an undiminished focus on how one can acquire true wealth.

To test his disciples, Dronacharya asked them, "The eye of the bird sitting on the tree is the target; can you see the bird?" Arjuna replied, "No, I can't see the bird; I can only see its eyes." Only with this kind of concentration and by remaining steadfast despite obstacles will one be able to achieve the target.

But what is that goal? To accumulate energy that can enable achieving the impossible—that is the goal.

An examination of life reveals that life flourishes only through an accumulation of energy. This energy is achieved only through one's own efforts. He who does not save, who depends on others, and who is a beggar, is dead even though he breathes.

He who has accumulated energy is strong, only he will distribute his accumulated wealth and enrich the world. Who will
take the path of this pursuit, this sadhana?

Only a few are invited to this, not for two years but for a life
of sadhana. Don't you see how the lives of so many are crushed
like worms, like specks of dust? Have you been afraid of the hec-
tic pace of the life cycle? Do you feel dejected for not under-
standing the ruthless and pilotless cause-and-effect relationship
of nature? But you have a divine vision in your heart; let it shine.
Maybe you will see a direction, a purpose in nature. You will see
that this world is alive, not just dead matter. Its diet is a meteor-
ite; its veins are flowing with molten metal. Not even a speck of
dust perishes, not even the tiniest energy perishes; perhaps life
is also imperishable. The highest form of excitement of life resides in its mental power. See, this holy country is enlivened
because of that energy, that shakti. Man reaches the same place
through service, devotion, and knowledge. You too choose one
of these paths. Let life and its consequences, this world and the
other world, be the goal of your pursuit. Throw your life in the
great war like a fearless hero.

12

The Injured Plant

The sky has been smoky in the West for a few years now. The eye has struggled to penetrate that darkness. Unclear—and unmanifested—cries have been suppressed by roaring cannons. But since the day the Sikh and Pathan, the Gorkha and Bengali, have gone to present their oblation in the great war, our powers of seeing and hearing have suddenly increased.

The final anguish of those whose lives have reddened the snowy white plains is hurting our hearts. What is this attraction that obliterates all gaps, that makes the near seem nearest, that makes us forgetful of whether they are strangers or our own? That attraction is compassion—it is only through the power of sympathy that real truth manifests itself. The ever-enduring plant world is standing still in front of us. Heat and cold, light and darkness, breeze and storm, life and death—they are playing a sport with the plants. They are being injured by many forces, but no sound of crying is raised. This very restrained, silent, and uncrying life has a shocking and poignant history, which I'll describe to you. If a man is injured, he screams. From this we deduce that he is hurt. The dumb does not shout; how will we know that he is injured? He becomes restless, gets the jitters, his hand and feet are excited; looking at him we can see that he has been hurt. Our compassion allows us to feel his pain. Causing injury to a frog doesn't make it scream, but it becomes restless, moving this way and that; but there is a great difference between man and frog. Whether a frog is injured or not, only the omniscient knows. Sympathy almost always looks upward, occasionally runs on the same place, and very infrequently is downward-looking. That the baser creatures experience happiness and sadness, respect and insult like us is something that some are suspicious of. There are no words about the low life. But that frogs experience something when they are hurt and that they respond to it must be accepted. They feel, they experience, they sense—I will use it in this sense. The human is hurt, the baser creatures respond—please do not object to these words. It is possible that our conditioning will make us see the agitated movements of a frog as suffering. Please treat this as metaphor. It is necessary to be careful about the use of words. A famous scholar from England has said that when the pearl is separated from the oyster, the latter feels no pain, rather its digestive canal is exhilarated by the warmth it experiences. Since no one has returned from the stomach of a tiger, the bliss of belonging to the stomach will remain outside description.

Life's Measuring Rod

Let us now see whether there is a measuring rod for the state of being alive. What is the difference between the living and the dead? Those who are alive respond when they are shaken. Not just that; those who are exceedingly alive, their response is capacious, great. The one who is almost dead, their response to being shaken is feeble. Those who are dead do not respond at all.

It is by causing injury that we can measure the living index. The high-spirited and full-blooded respond fully to injury. And those who are weak, even a lot of chasing does not lead to any answer from them. Imagine that somehow my finger is wounded repeatedly. The injured finger, being excited, is therefore moving. A mild injury will cause it to move a little, a great blow will make it move wildly. The naked eye cannot measure this difference. It is necessary to calibrate this excitement in writing. The experiment that I'll show a little later will give a sense of such an instrument. With a little excitement after a mild injury, the pen moves upward the lines denoting excitation having little volume. For an injury of greater impact, the line is bolder.

Not only that. We regain our composure after the unexpected attack; the contracted finger gradually returns to its normal position. When struck, the line drawn by the pressure of the contracted finger suddenly moves upward. It takes some time to become calm again; the elevated line gradually returns to its original state. The pain from the strike reaches its full intensity quickly but takes time to subside. Similarly, the response from the contraction occurs quickly; the line of extension indicating recovery takes longer. A severe strike results in a greater response, and it takes longer to recover. The pain also lasts longer. If the living muscle remains in the same condition and is struck repeatedly, the responses are similar. A living muscle is, however, never in the same condition at all times; because the external world and past history are constantly shaping us, our nature too is changing moment to moment. Sometimes cheerful, sometimes depressed, sometimes dying. These internal changes are not often visible externally. Someone who appears good-natured might have a prickly temper that flares up at the slightest provocation; another person might be unresponsive to everything. There are personal differences and situational changes, and, besides these, there are the histories and memories of a life that leave invisible marks. Will those hidden stories ever be revealed? At first it seems this is entirely impossible. Let's see if the impossible can become possible. How can we assess someone's nature? What is the difference between the fake and the genuine? To test a currency note, you need to strike it and listen to the sound of its response. The responses of the genuine and the fake notes are entirely different; one has a tune, the other is completely out of tune. A person's nature can also be tested by striking it.

Fate tests humans with harsh blows; the test of the genuine and the fake happens only at that time.

Perhaps this is how the nature and history of a living being can be uncovered—by striking it and recording its response. The response script is just a line; some lines are a little longer, some a bit shorter. How can such inexpressible, intimate, and mysterious histories be revealed in the slight differences between two lines? It seems impossible, but it is not as improbable as it appears. Due to the fault in our stars we might one day have to appear in court as defendants. There the defendant does not have the right to make grand speeches. To the prosecutor's interrogation one can only answer yes or no, we can only respond in two ways—by moving the head up and down or side to side. If the ink is applied to the defendant's nose and a piece of white blotter paper is held in front, two types of responses will be recorded on the paper. This will be the true mark of the nose, and from these two lined responses the righteous judge will examine our entire life. The result of that judgment will determine our future residence—in Calcutta or Andaman, in this world or the next.

So far I've spoken about the human. Now let me speak about plants and their hidden histories. To examine a plant, it must be stimulated by some specific kind of impact, and the response it gives has to be recorded by the plant itself. From the style of that writing, its present and past history must be revealed. Therefore, to make this difficult endeavor successful, we need to observe the following:

1. What kinds of impact excite a plant, and how can the degree of those impacts be measured?
2. How does the tree emit a signal in response to those impacts?
3. How can those signals be recorded in writing?
4. How can the plant's history be uncovered from the style of that writing?
5. How does a plant feel when its branches, its equivalent of arms, are cut?

Excitation in Plants

As I've said before, when any part of our body is injured, it produces a sensation of disturbance that causes that part of the body to contract. A shock caused by the contraction from the injured area travels through the nerves and hits the brain—this we then, depending on the nature and intensity of the injury, perceive as either pleasure or pain. Even if the limbs are all tied up and their movement thereby restricted, the transmission of signals through the nerves does not stop. If a plant is connected to an electrical apparatus, it is observed that as soon as the tree is struck, it gives an electrical response. No such response is observed after the plant's death. This is how I have been able to prove that all types of plants and their various parts feel hurt and injured when they are struck.

There are certain plants that respond by moving—the touch-me-not plant, for instance. At the base of each leaf, the plant muscle is relatively thick. Just as our muscles contract when injured, the plant muscles at the base of the leaf contract similarly when struck, causing the leaf to droop or fall. After the sudden contraction due to the injury, the plant recovers, and the leaf returns to its original position. Just as humans respond by moving their hands, the touch-me-not plant responds by moving its leaves.

Just as humans can be stimulated, the touch-me-not plant can be stimulated in the same way—by striking it with a stick, by pinching it, scorching it with a hot iron, or burning it with acid. The leaves respond to these stimuli. These severe stimulations cannot be endured by the leaves for long, and the plant eventually dies. For prolonged experiments, therefore, a gentle method of stimulation is necessary, one that ensures that leaves do not perish and that the degree of stimulation remains consistent.

The plant must be awakened from a dormant or stationary state with ease, gently. The princess in the fairy tale was put to sleep by magic; her sleep was broken by the touch of a gold or silver wand. From the preceding experiment it can be understood that the touch of a gold or silver wand caused the touch-me-not plant and a motionless frog to respond by moving their leaves and bodies. This is because, upon contact with two different metals, an electric current flows, and this electric force excites both living beings and plants in the same manner. The advantage of using electricity for excitation is that its strength can be increased, decreased, or kept constant. By adjusting the machine's dial, the electric shock can be made as fierce as lightning to instantly destroy life, or it can be made progressively milder. Such gentle stimulation does not harm a tree.

The Writing Instrument of Plants

I have spoken about the responses of plants. Now, the difficult problem is how to record the plant's responses. In animals, responses are usually recorded by attaching a pen. However, just as tying a fan-like basket to a sparrow's tail does not help it to fly, attaching a pen to a plant leaf does not help it to write. In fact, the tiny leaves of the telegraph plant [*Desmodium* sp.] cannot even bear the weight of a thread, so there was no possibility that it could push a pen to write a response. I therefore adopted a different approach. Light rays have no weight. Initially I manually recorded various script styles of the tree leaves using reflected light rays. This task took many years to complete. When I presented these novel findings to biologists, they were completely astonished. They eventually informed me that these findings were so unprecedented that they would only accept them if, one day, the tree itself could write and provide testimony.

The day I received this response, it felt as if all light had disappeared from my eyes. I had always known that success was just the flip side of failure. I tried to understand this anew. Twelve years later, the curse became a blessing. I will briefly recount those twelve years. I entirely redesigned the apparatus. I created a lightweight pen using the finest wire. This pen was also mounted on a jewel made of emerald so that even the slightest movement of the leaf could easily turn the pen. After many days, the pen began to vibrate in response to the movements of the tree leaf. But the pen could not overcome the friction against the paper. So I applied lamp black onto smooth glass instead of paper. White writing appeared on the black writing surface. This reduced the friction considerably, but even so, the tree leaf could not push the pen through the slight friction barrier. It took another five or six years to make the impossible possible. This was achieved through the invention of my "resonant" device. I won't test your patience by describing the construction details of these apparatuses.

However, it is necessary to mention that these devices record various responses of the tree. The growth of the tree is measured moment by moment, its autonomous responses are recorded, and the lines of life and death determine its lifespan.

Decoding the Internal History of Plants from Their Writing

Interpreting the writing patterns of plants is time-consuming. In an excited state, the responses are larger; in a despondent state, the responses are smaller; and in a near-death state, the responses are almost nonexistent. The response records you see before you were made when the sky was filled with abundant light and the tree was in a cheerful state.

That is why the responses are so large! Suddenly, for some unknown reason, the magnitude of the responses became smaller. If any change occurred in the meantime, it was beyond my perception. When I went outside, I saw a small cloud drifting across the sun. The slight reduction in sunlight was not noticeable from inside the room, but the tree sensed it and expressed its melancholy with a small response. As soon as the cloud passed, it responded with its previous cheerfulness. I've said before that I had proven through electrical experiments that all trees have sensory capabilities. Scientists in the West were unable to believe this for a long time. Recently, a date palm tree in Faridpur proved my point. This tree would raise its head at dawn and bow its head to touch the ground at dusk. I have been able to prove that this was due to the tree's sensitivity to external changes. From the examples given, you can understand that a tree's responses can reveal its hidden life history. Many new facts about life have been discovered through experiments on plants. It seems that not only scientific truths but also many philosophical questions may be resolved through these findings.

Container Oil

I hear that the debate over whether a dog wags its tail or the tail wags the dog has not yet been resolved. Some say the dog wags its tail; others say the tail wags the dog. Similarly, does the leaf move, or does the plant move? Is it the container of oil or the oil in the container? Who initiates the movement, and who responds? In England there is much discussion about Indian society. It is said that women here cannot act on their own will; they merely move like puppets at the behest of men. Who follows whose lead? Who holds the reins? Who moves, the dog or its tail? The experiences of those involved suggest otherwise.

Despite outward appearances of dominance and swagger, it is said that all this puppet dancing is directed from some inner chambers. There are times when a woman herself cuts the binding string. She commands the one she has so far protected with her sari, "Go away, go afar, I carry only my blessings, I welcome you into the hands of death!"

When the touch-me-not plant is struck, its leaves fall. Is it the leaves that move or the tree? This can be tested. First, if the tree is held still, it cannot move; only the leaves can. But if the leaves are held and the roots are lifted from the ground, it is seen that the plant moves in response to the strike while the leaves remain still. When a part of the body is struck, the pain from that strike spreads through all the veins and branches of the tree, and one part's misfortune is shared by all. Although the tree is composed of hundreds of thousands of leaves and branches, a single connection binds them together. It is this unity that allows the tree to stand tall and strong against external storms and blows.

Response of the Injured

Let us now see the various ways an injured tree manifests its distress outwardly. I will describe two types of responses regarding this matter in trees. I will explain whether the growth rate of a growing tree increases or decreases when it is struck by a knife. I will then show how a tree and its severed leaves react to the injury when the leaves are cut off. To understand the natural growth rate of a plant takes a lot of time. The growth rate of a tree is six thousand times slower than that of a snail. I had to therefore invent a new instrument called the crescograph. It records the growth rate by amplifying it one crore times. Where the microscope fails, the crescograph is a million times more sensitive.

You may not be able to grasp the concept of a crore-fold increase, so I'll give an example in the form of a story. Once, there was a race between the Bengal-Nagpur Railway and the East India Railway to see which could go faster. A snail watching this couldn't contain its laughter. It immediately climbed onto the crescograph. After a while, when it looked back, it saw the trains had fallen far behind.

I wanted to name the device Briddhiman ("Increasing") instead of Crescograph. But it didn't happen. Initially, I gave Sanskrit names such as Kunchanman and Shoshanman to my new instruments. Trying to promote nationalism proved to be quite challenging, however. These names were so peculiar that British newspapers mocked them. Only the leading newspaper in Boston supported me. The editor wrote, "The right to name an invention belongs to the inventor. New devices are typically named in ancient languages like Latin and Greek. If that is the case, why not from the ancient yet living Sanskrit?" I tried to push the names forcefully, but the outcome was different. During a lecture at an American university, a renowned professor asked me to explain my device, "Kanchanman." At first I did not understand; eventually, I realized that Kunchanman had been transformed into "Kanchanman." Following Mr. Hunter's method of spelling, I had spelled it "Kunchan," but it became "Kanchan."

A special feature of the Roman alphabet is that a vowel can be pronounced from "a" to "au" arbitrarily, except for "ri" and "li." Even these can be managed with dots above or below.

Whatever the case may be, I realized that it might be possible to make Hiranyakashipu chant the name of Hari, but it is entirely impossible to make an Englishman speak Bengali or Sanskrit. This is why our Hari must become Harry. After seeing all this, my desire to name the instrument Briddhiman completely vanished. *Buddhiman*—"intelligent"—would become Burdwan. Crescograph is much better in comparison.

This instrument records how much a tree grows every second. From this, it was known that this particular plant was growing at a rate of forty-two millionths of an inch per minute. I hit the plant with a cane. The plant's growth reduced immediately. It took the plant more than half an hour to forget this injury. After that it began growing spontaneously as before. O cane-wielding schoolmaster, there is no doubt that some have become judges of the High Court because of being pulled by the ears by you. That boys would grow tall because of being struck by your cane is doubtful though. All kinds of injury stunt growth. When pricked with a needle, the plant's growth reduced to a quarter of its usual rate. Even after an hour, it had not recovered from the injury, with its growth rate still less than half. A longitudinal cut with a knife inflicted an even more severe injury. This caused the growth to stop for a considerable period of time. A zigzag cut was even more dire than a straight cut. Housewives should remember this when cutting a koi fish.

The Loss of Sensitivity Due to Injury

After this, I cut the leaf of a touch-me-not plant. All the leaves, both the cut leaf and the leaves on the plant, drooped down. It was necessary to observe the condition of the cut leaf and the injured plant. Upon examination I found that both remained completely unconscious for three to four hours. The subsequent events—its history—were quite extraordinary. I provided the cut leaf with nourishing sap to keep it alive. After four hours, the leaf perked up and gave a strong response. Its attitude seemed to be: "What happened? This is good! I was tied to the plant for so long, and now my body feels so light!" The leaf continued to respond stubbornly in this manner throughout the day. However, the next day, for some unknown reason, its responses significantly diminished. After fifty hours, the leaf collapsed face down. Soon after, it was dead.

The story of the plant from which the leaf was cut was different. It slowly began to recover. It did not exhibit a "nothing matters" attitude. It had to make do with what it had. Gradually, the injured plant managed to cope with its pain. It shook off the temporary weakness and was able to respond as it did before.

Homeland

Why, then, this difference? What causes the branchless tree, though injured and near death, to survive after some time, while the detached leaf, despite being nurtured in various ways, succumbs to death? It is because the tree's roots are firmly established in a specific ground, nourished by the essence of that land. That ground is its *swadesh*, its homeland and its patron.

There is also an inherent strength within the tree that has protected it from annihilation through the ages. Many changes have occurred externally, but it has not been defeated by fate. In response to external blows, it has battled these changes with the fullness of its life. It has accepted the necessary changes and discarded the unnecessary ones, like withered leaves. It has survived external threats.

All along, it has had another strength. The memory of having originated from the seed of the banyan tree is imprinted on every part of it. Because of this, its roots are firmly established in its mainland, its head reaches upward in search of light, and its limbs spread out to provide shade. With what strength does it survive despite its injuries? Through patience and steadfastness it tightly embraces its place, balances the inner and outer harmonies through perception, and retains the accumulated strength of many lifetimes through memory. And what about the unfortunate one that detaches itself from its birthplace and homeland, nurtured by foreign sustenance, and forgets its national memory? What strength must such an unfortunate being possess in order to survive? Destruction stares it in the face, annihilation is its consequence.

(Lecture at Sahitya Parishad, 1919)

13

The Flow of Stimulus in the Nervous System

How does news of the world reach our inside? Our external senses extend in all directions. A push or injury to them is transmitted as news inside. The waves in the sky hurt the eye—that news we have come to understand as light. The waves in the air that injure the ears—that news we experience as sound. If this injury is mild, it mostly feels pleasurable. But if the degree of injury is increased, the feeling is different. Mild touch brings pleasure, but an injury that stuns cannot bring any joy.

Electric waves flow from place to place through the telegraph—that is how it reaches distant lands, through signals. If the wires are cut, the news will stop. The same electric flow generates different kinds of signals in different instruments—it moves a needle, rings a bell, or lights a lamp. The different stimuli that excite or hurt the nervous system, those we experience differently—sometimes as sound, sometimes as light, sometimes as touch. If the stimulus passes through the muscles, they contract. Just as the severing of wires leads to a cessation in the flow of news, so information from outside stops coming inside if there is a cut in the nervous system.

Self-Vibration and Internal Power

I have told you about the response caused by external injury. There is another kind of response that happens on its own. That self-vibration or excitement of the pulse is caused by some unknown power. The rhythmic vibration in our heart is an example of this. This happens completely on its own. We can see examples of this in the plant world. The two tiny leaves of the telegraph plant, *Desmodium motorium,* move on their own. A distinguishing characteristic of this internal vibration is that it is not affected by any external force; on the contrary, it resists the external force. So you see that two kinds of forces excite living beings—external force and internal force. Usually the internal force resists the external force.

How Can What Is Unacceptable to the Senses Be Made Acceptable?

It is the degree of injury that determines the increase and decrease of stimulus or excitement. There are many things that are happening about which our senses are not bothered. When light grows weak and then feebler, then the visible becomes invisible. Even though the eyes are being injured by light, the stimulus is so feeble that it cannot travel for any great length through the nervous system—that is why it cannot wake up the senses. Will what is unacceptable to the senses ever become acceptable? What I had discovered for a moment I'm not able to see anymore. How, then, will my vision become sharper and my senses grow stronger?

On the other hand, how will the senses, battered by injury from the external world, be soothed? O Timid One, even though you will die one day, you are experiencing the pain of untimely death even before you have died. Even though you are incapable of preventing attacks from the external world, you are the master of your inner world. The path through which news from the outside world comes to you, will it ever obey your orders, will it expand or contract according to your instructions?

Sometimes things such as these are known to have happened. What a distracted state of mind did not allow me to see or hear, by disciplining my mind I've been able to see and hear. In such a situation, it seems that will and conditioning might increase the capacity for feeling. When news of the world reaches the inside through the nervous system, does the nervous system adjust to open a half-closed door freely? There are perhaps other ways by which fully opened doors close completely.

Resistance to External Pressure

I saw something like this when I was living in Kumaon. A tiger from the Terai region was destroying everything. Within the span of just a few days more than a hundred people had been eaten by the tiger. The government had tried many measures to kill the tiger, but these had all come to naught. Helpless, the villagers went to Kalu Singha for help. He used to be a hunter a long time ago, but since the restrictions on arms he hadn't used his single-barrel gun in a really long time. The tiger had killed a buffalo in the field during the day; I had heard the wailing of the buffalo clearly. The tiger would return to the field at night—and with that hope in mind, Kalu Singha was waiting for it in his hideout. Evening came, and the tiger appeared, like Yama, the god of death himself, only three hands—one foot—the distance between them. Kalu Singha's body was shaking in fear; he wasn't being able to hold the gun at all. As Kalu Singha told me later— "Then I scolded myself and said, 'What is this, Kalu Singh? I've sent you to save the lives of wives and sisters and children, and you are lying down hiding behind the bush?' Immediately something ran out of me like fire, that made my body hard as iron. Then I went and stood in front of the tiger. The tiger leapt at me, the sound from my gun erupted right at the moment, and the tiger was killed."

Something runs through the nerves that turns bodies into something hard as iron. Then no fear can penetrate through an iron armor. What changes occur in the nervous system to make this impossible possible? The stimulus and excitement in the nervous system are invisible; its nature, the rules that govern it, we know nothing about it. Believing that scientific investigation would reveal the reason and theory behind this, I have spent twenty years of my life pursuing the answers.

The Nervous System in Plants

Before everything else, I investigated plant life. Pfeiffer, Haberlandt, and other European experts had concluded that plants do not have a nervous system; why, then, does pinching the touch-me-not plant cause its leaves to fall off? In response, they said that pinching a plant causes it to produce a flow of water the force of which leads to its leaves falling off. That this disposition is delusional has been proved by my experiments. Firstly, without pinching the touch-me-not, the plant can be stimulated without the production of water at all. It has also been seen that the characteristics one notices in animals, in organisms, one finds in plants as well. The velocity of water inside a pipe is not dependent on its temperature, on whether it is warm or cold; but the excitement in the nerves is doubled with an increase of nine degrees centigrade. Exactly the same thing happens in plants. Very cold temperatures render the plant's nerves passive; then the stimulus cannot flow through it at all. The application of chloroform suspends the flow of stimulus. That plants have a nervous system—my inference has now been accepted everywhere.

The Increase and Decrease in Stimulus and Its Molecular Disposition

Let us first see how excitement in the nerves is transmitted. Once there is a clear idea about this, we will later be able to see how this stimulus is accelerated or alleviated. The nervous system is composed of many molecules; under normal circumstances, every molecule stays in its place. But when they are injured, they sway this way and that; in this swaying to and fro is the state of excitement. When a molecule vibrates, its neighboring molecules begin vibrating as well, and that is how the excitement reaches from one end to the other. I can imagine how the vibration of the injured molecule travels long distances. Think of a table with a pile of books arranged neatly on it. If the book on the extreme right is pushed to the left, the book on top will fall on the second book and both those on the third. This is how the push caused by the injury will travel to the other side.

The books were in one orderly pile, and to move the book on the top from its position requires some strength; imagine that the measure of that strength is 5. If the strength of the push was 3 instead of 5, then the books wouldn't have toppled over, and the neighboring books would remain still as well. That is why when the pressure on the external senses is very weak, the stimulus cannot travel very far, and so the hurt from the external world does not affect the senses. Imagine that the books are not in a straight pile but leaning against each other toward the left. Now even a little push will cause the books to topple over and the push to travel from one end to the other. Previously, if the measure of the push had been 3 and not 5, it would not have been possible to move the first book; now it will move easily. If the books were pushed in the opposite direction, the first book will not be dislodged. The push will not travel far; it is as if the path to the destination has been completely closed. From this example we understand that the molecules of the nervous system can be arranged in two ways. When facing toward, the stimulus will be communicated through the senses. When averse, or facing in the other direction, a strong push will not be communicated to the inside.

Experiments

How the problem related to the moderation of excitement can be solved I have described in a broad manner. Whatever thoughts I've had about it are subject to the results of experiments. But in what way can molecular composition be positive-facing or negative? It is seen that electrical flow turns magnetic needles unidirectional; when there's flow in the opposite direction, the needles turn in the other direction. If electric current is passed through water, the molecules are agitated and their disposition becomes inclined toward the flow of current.

This can be seen working in two different ways in the nervous system. My first experiment was performed on the touch-me-not plant. The hurt I inflicted on it was so feeble that the *Mimosa pudica* was unable to feel it. Then the molecular formation was turned forward-facing. The hurt that the touch-me-not hadn't felt before it now felt immediately, and it showed this by moving its leaves emphatically. After that I turned the molecular disposition in the opposite direction. Now when I hit the touch-me-not with a lot of force, the plant remained indifferent to it, the leaves showed that they were ignoring it by not responding at all.

I began conducting experiments on frogs in the same manner. The injury that the frog had never experienced in its life it began to feel when electric current was passed through its nervous system in the forward-facing molecular disposition: the frog responded by shaking its body. After that I "rubbed salt in the wound." The frog began to grow jittery. But as soon as I changed the direction of the molecular insertion, it was as if the painful flow stopped midway and the frog fell completely quiet.

So we see that excitement in the nerves can be increased or decreased as one wishes. This increase or reduction depends on the molecular insertion or molecular signaling. In one form, the excitement increases; in another, the stimulus falls numb. It has also been seen that the excitement generated by this molecular insertion can be controlled through an external force. This is not an unexpected or divine event but a verifiable scientific truth. The relation between cause and effect is irrefutable.

What happens because of an external force can also take place because of an internal force. Just as an external injury leads to the contraction of muscles in the arm, an internal wish can also lead to the contraction of the hand. An instruction that is the opposite of this might turn the arm sloth-like. Here we see that internal desire can control molecular insertion in the nervous system. When that happens, internal force can control the excitement in the nervous system as well. But the power to control the two kinds of molecular insertion needs years of practice and dedication. An infant cannot walk in the beginning. But days of trying soon makes walking a natural habit.

Man is therefore not just a slave of the invisible; in them is the power to neutralize the external world. It is according to their wish that the inside-outside gate is sometimes unlocked, sometimes closed. This is how it emerges victorious over physical and mental weakness. The feeble message that wasn't audible it is now able to hear; the target that wasn't visible is now flaming. It is through these other ways that all the world's horror will now be a thing of the past. In the kingdom inside, even when there's a storm outside, the internal voluntary power will keep it unruffled.

Inside and Outside

The power inside is voluntary. From which level of life did this strength originate? Dry grass flows away in water currents. But organisms are not controlled only by the flow of the world outside; rather, stimulated by the injury from the wave, it fights against the flow. At which level does it gain the strength to do this? The tiniest organism accepts the pressure from the outer world sometimes; sometimes, with the help of its internal strength, it rejects this pressure. The ability to accept or reject is, after all, the power of wishes and desire. And how did the internal strength come to be? Is internal strength completely different from external strength? I've told you before that it is because of the internal strength of the *Desmodium motorium* that its leaves keep moving on their own. But keep the plant in the dark for two days and its leaves go completely lifeless. This is because the strength that had accumulated inside has completely exhausted itself. If the leaves are exposed to a little light at that moment, it will be seen that the leaves respond with a little movement; but taking away the source of light causes the vibration of the leaves to stop immediately.

When they are exposed to light for a long period of time, a strange thing will be noticed. Even after the light is turned off, the two tiny leaves continue to move for a significant period of time. What can be more astonishing than this? It is seen that light, which was an external force, has been accepted by the plant as its own and that energy that had accumulated from outside has been turned into a form of internal energy. External and internal strength are, by nature, the same; the difference is that what was on the other side of the curtain has now come on this side; what was a stranger has now become one's own. You'll also see that these self-vibrating leaves are not agitated by an external force. Now that the external force has been neutralized, its internal strength has made it capable of protecting itself from the outside. When its internal energy is exhausted, only then will it respond to the external world and will later reject the external stimulus again. At what level or stage does internal strength originate voluntarily?

At the time of birth, I was tiny and helpless, thrown into a sea of force and power. Then these forces from the external world entered me and nurtured my body into growth. Mother's milk, along with her affection and care, entered me, as did the love of friends, to cheer me through this life. The injuries of difficult times generated internal strength that has accumulated inside to help me deal with the external world now.

What is my own contribution in this, then? Who is responsible for this—you or me?

Through the excitement of one life, you have fulfilled another life. Many, in obedience to your instructions, have gone out in search of knowledge, and for the good of man have sacrificed the riches bestowed on them by the state for a life of poverty and suffering, in service to their country. The dispersed strength of those lives has brought wisdom and dharma to some, and valor and prowess to others.

In the tussle between external and internal forces, life is manifesting itself clearly. The root to both is the same great power, because of whom the animate and the inanimate, the molecule and the universe, are inspired. Life is an expression of that power and its excitement. It is that power that will make man reject the devil and be promoted to god.

14

Hajir!

Suddenly someone answered, shouting—"Hajir!" I hadn't heard anyone call for me, so I answered in a rather doleful but devoted and attentive voice—"Yes, my lord?" Who is your lord, whose instructions have brightened and energized you?

How astonishing! Just one word has excited all the spheres of life. Dormant memories are awakened today—what was outside the sphere of hearing is now audible; what was outside intelligence has now become meaningful.

Now I am able to understand that it is not only from outside that orders come—they come from inside too. I used to think that everything has happened because of my wishes. Am I one? When my mind is still, I am able to hear the conversation between two selves. They are the ones running me, making me work. Among them, the bad self is me; who is the good, intelligent self then?

Events from twenty-seven years ago return to me in relation to this. I hadn't ever been able to write it, but someone inside me made me start. It was in obedience to their order that I wrote "Akash Spandan" and "Adrishya Alok" ["The Resonance and Conceivable World of the Sky" and "Invisible Light"]; then they made me write "Udbhid Jeebon Manobiyo Jeeboneri Chhaya Matro" ["Plant Life Is a Mere Shadow of Human Life"]. Earlier I didn't know very much about life. Whose orders made me write like this? There was no relief or getting away even after writing this; the critic inside began saying—"All the things you have written, have you tested them—which of them are true and which of them false?" I replied, "How can I decide this about subjects on which even experts have failed? They have numerous instruments and workshops, there's nothing here, how will I make the impossible possible?" Even this did not stop the critic. I was thus compelled to get a carpenter to create an instrument within three months. The findings and results that it produced did not surprise just me but caused astonishment to scientists abroad as well.

I found fame very soon and was invited for a ceremony of commendation abroad. William Ramsay, the famous scientist, paid me many compliments; he later said, "Some people might feel that from now on a new era of knowledge will begin in India; but just one cuckoo call should not be mistaken for the arrival of spring." My bad self must have taken over my being that day, for my answer was audacious. I said—"All of you do not need to worry; I am certain that in the scientific sphere a hundred cuckoos will soon herald spring in India." Now the time has come— what I had mistakenly thought to be a wrong inclination I now see as the right choice. Five years since that blessed moment, one day brighter than the other, and all paths begin to open.

It was at this time that the order came to give up the easy path for the uncertain and inaccessible road. I was experimenting with the wireless at that time. I had begun to see that the initial response of the instruments was immense, then it would gradually go weak before disappearing. I had written this in a scientific research journal—"It is best to begin the experiments at daybreak; this is because the instrument gets tired by day's end." Immediately the critic inside cried out—"Is the instrument a human, that it should feel exhausted?"

Why does the instrument feel tired? It became impossible for me to avoid this question. There were many inventions that were waiting to be written down. Leaving them aside, I had to turn to investigate these new questions. I gradually began to see that even lifeless metal gets excited and exhausted. If the excitement is suspended for some time, the tiredness goes away. I saw this behavior even more clearly in plants. This made me aware of the unity in diversity.

I had thought that I would put these discoveries and inferences in the hands of theorists of life science for further exploration and return to the physical sciences; but the opposite happened—things took an unexpected turn. I had shared all the experiments and inferences with the Royal Society. Burdon-Sanderson, the best-known scientist in this field, said, "What you have said about the life sciences we have tried to prove without success in the past. So your words are impossible and inconsequential. Your entry into this field has been illegitimate. You have found fame in the physical sciences, and there are many laurels awaiting you on this path. Recede from the path that you are ignorant of." Provoked by my audacious self, I replied—"I will not recede. This friend's path is mine. I leave the easy path from today. What has been rejected today is the truth." Whether willingly or not, everyone will have to accept that.

The fruit of this mischievous self did not take long to ripen. Paths in all directions were closed to me, and all the light was suddenly extinguished. But soon afterward the feeble light inside brightened. What I hadn't been able to see in the harsh light I could see now. In this way, between hope and hopelessness, twenty years passed.

One year ago, suddenly, I could hear the orders: "Go abroad. Foreign travel!" Who will listen to me there? Again I heard a stubborn and sharp voice, "My name is Order, yours is Tamil! Who are you to think of gains and profits?" I had to obey the order.

Then all the closed doors in all directions opened. On whose instructions did this happen? Was this a dream? Those who were opponents were now dear friends. What had been rejected was now common acceptance. Twenty years ago, what I had considered to be wrong and arrogant, I could now see as the right self.

That is why I do not know what is audacious and what is right and intelligent. What is big and what small, the mind doesn't understand. Forgetting the success of the good times, it is the hardships and failures of the difficult times that are coming to me. At that time, I was rejected by everybody; the affection of only a couple of people kept me alive. They are now behind the curtains of darkness. Can unexpressed tears reach there?

When life was at its full strength, I could not hear your orders because of the noise around me. Now I can; but all my strength is gradually growing numb. One day, the curtain will descend on your instructions, what had been created from the earth will fall as dust. What will they carry when they reach you at last? Little are their achievements, numerous their offences. But who is there to say? What is good intention, and what is wrong and audacious—in solving this puzzle, life has gone by. When there is no explanation to give, then they will prostrate themselves at your feet and only say—"Asami hajir!": the accused is present.

Index